Advances in Energy Materials

Advances in Energy Materials

Ceramic Transactions, Volume 205

A Collection of Papers Presented at the 2008 Materials Science and Technology Conference (MS&T08)
October 5–9, 2008
Pittsburgh, Pennsylvania

Edited by

Fatih Dogan
Navin Manjooran

A John Wiley & Sons, Inc., Publication

Published by John Wiley & Sons, Inc., Hoboken, New Jersey.
Published simultaneously in Canada.

Library of Congress Cataloging-in-Publication Data is available.

ISBN 978-0-470-40843-8

Printed in the United States of America.

10 9 8 7 6 5 4 3 2 1

Contents

Preface vii

INDUSTRIAL PERSPECTIVE OVERVIEW

The Role of Materials and Manufacturing Technologies as Enablers in Gas Turbine Cooling for High Performance Engines 3
Ron S. Bunker

ENERGY MATERIALS

Synthesis, Sintering and Dielectric Properties of Nano Structured High Purity Titanium Dioxide 23
Sheng Chao and Fatih Dogan

Sorption/Desorption Properties of MgH_2-Oxide Composite Prepared by Ultra High-Energy Planetary Ball Milling 31
Y. Kodera, N. Yamasaki, J. Miki, M. Ohyanagi, S. Shiozaki, S. Fukui, J. Yin, and T. Fukui

Ab Initio Study of the Influence of Pressure on the Hydrogen Diffusion Behavior in Zirconium Hydrogen Solid Solution 41
Y. Endo, M. Ito, H. Muta, K. Kurosaki, M. Uno, and S. Yamanaka

EBSP Study of Hydride Precipitation Behavior in Zr-Nb Alloys 51
Shunichiro Nishioka, Masato Ito, Hiroaki Muta, Masayoshi Uno, and ShinsukeYamanaka

FEM Study of Delayed Hydride Cracking in Zirconium Alloy Fuel Cladding 59
Masayoshi Uno, Masato Ito, Hiroaki Muta, Ken Kurosaki, and Shinsuke Yamanaka

The Effect of Manganese Stoichiometry on the Curie Temperature of $La_{0.67}Ca_{0.26}Sr_{0.07}Mn_{1+x}O_3$ Used in Magnetic Refrigeration 71
Biering, M. Menon, and N. Pryds

Preparation of Electrocatalytically Active RuO_2/Ti Electrodes by Pechini Method 77
O. Kahvecioglu and S. Timur

The Myriad Structures of Liquid Water: Introduction to the Essential Materials Science 87
Rustum Roy and Manju L. Rao

Preparation of $CuInS_2$ Films by Electrodeposition: Effect of Metal Element Addition to Electrolyte Bath 99
Tomoya Honjo, Masayoshi Uno, and Shinsuke Yamanaka

Preparation of High-*Jc* MOD-YBCO Films for Fault Current Limiters 109
M. Sohma, W. Kondo, K. Tsukada, I. Yamaguchi, T. Kumagai, T. Manabe, K. Arai, and H. Yamasaki

NANOTECHNOLOGY FOR POWER GENERATION

Modeling of Electromagnetic Wave Propagation of Nano-Structured Fibers for Sensor Applications 117
Neal T. Pfeiffenberger and Gary R. Pickrell

Increased Functionality of Novel Nano-Porous Fiber Optic Structures through Electroless Copper Deposition and Quantum Dot Solutions 123
Michael G. Wooddell, Gary Pickrell, and Brian Scott

Thermopower Measurements in 1-D Semiconductor Systems 135
Sezhian Annamalai, Jugdersuren Battogtokh, Rudra Bhatta, Ian L. Pegg and Biprodas Dutta

Structural Changes and Stability of Pore Morphologies of a Porous Glass at Elevated Temperatures 145
Brian Scott and Gary Pickrell

Author Index 159

Preface

Increasing awareness of environmental factors and limited energy resources have led to a profound evolution in the way we view the generation and supply of energy. Although fossil and nuclear sources will remain the most important energy provider for many more years, flexible technological solutions that involve alternative means of energy supply and storage need to be developed urgently.

The search for cleaner, cheaper, smaller and more efficient energy technologies has been driven by recent technology advancements particularly in the field of materials science and engineering. This volume documents a special collection of articles from a select group of invited prominent scientists from academia, national laboratories and industry who presented their work at the symposia on Energy Materials and Nanotechnology for Power Generation at the 2008 Materials Science and Technology (MS&T'08) conference held in Pittsburgh, PA, from 5th -9th October These articles represent a summary of the presentations focusing on both the scientific and technological aspects of energy storage, nuclear materials, nano-based sensors, catalysts and devices for applications in power generation, solar energy materials, superconductors and more.

The success of the symposia could not have been possible without the support of staff at The American Ceramic Society and the other symposia co-organizers Drs. Masanobu Awano, Wayne Huebner, Dileep Singh, and Gary Pickrell. The organizers are grateful to all participants and session chairs for their time and effort, to authors for their timely submissions and revisions of the manuscripts, and to reviewers for their valuable comments and suggestions.

Special thanks to Dr. Ronald Bunker from GE Global Research Center who kindly provided the introductory article highlighting the industry perspective towards the need for advanced energy materials.

Fatih Dogan
Missouri University of Science and Technology

Navin Manjooran
Siemens AG

Industrial Perspective Overview

THE ROLE OF MATERIALS AND MANUFACTURING TECHNOLOGIES AS ENABLERS IN GAS TURBINE COOLING FOR HIGH PERFORMANCE ENGINES

Ron S. Bunker
GE Global Research Center
Niskayuna, New York, USA

ABSTRACT

Gas turbines contribute a significant portion of the world's power demand, as well as the majority of aircraft propulsion needs. Today's complex cooled gas turbine components would not be possible without continuous advances in both materials and manufacturing science and technology. The turbine hot gas path is the most costly portion of the engine and as a consequence, improvements in materials and manufacturing, especially those that enable better cooling, carry a high return on investment. This summary identifies the elements in turbine cooling that can benefit from materials and manufacturing enabling technologies, as well as many of the emerging means for making these a reality.

INTRODUCTION

The gas turbine is a specialized engine designed to convert chemical energy into one or more useful forms of energy, such as thrust, shaft work, and process heat. Gas turbine engines for aviation and marine propulsion, power generation, and combined heat / power applications are most commonly in the form of continuously rotating axial turbomachinery. An overall engine schematic is shown in Figure 1 for the CFM56-5B commercial aviation gas turbine engine. As a thermodynamic Brayton cycle, the efficiency of the gas turbine engine can be raised substantially by increasing the firing temperature of the turbine. Modern gas turbine systems are fired at temperatures far in excess of the material melting temperature limits. This is made possible by the aggressive cooling of the hot gas path (HGP) components, the use of advanced materials for structural components and protective coatings, the application of high efficiency aerodynamics, the use of prognostic and health monitoring systems, and the continuous development of improved mechanical stress, lifing, and systems interactions and behavioral modeling.

The high-pressure turbine (HPT) section of the engine, shown in Figure 2, encompasses all of these challenges simultaneously. For example, the technology of cooling gas turbine components via internal convective flows of single-phase gases has developed over the years from simple smooth cooling passages to very complex geometries involving many differing surfaces, architectures, and fluid-surface interactions[1]. The fundamental aim of this technology area is to obtain the highest overall cooling effectiveness with the lowest possible penalty on the thermodynamic cycle performance. Figure 3 provides a generic view of the gross cooling effectiveness for turbine airfoils with the cooling technologies developed over the years. The use of 20 to 30% of the compressor air to cool the HPT presents a severe penalty on the thermodynamic efficiency unless the firing temperature is sufficiently high for the gains to outweigh the losses. In all properly operating cooled turbine systems, the efficiency gain is significant enough to justify the added complexity and cost of the cooling technologies employed. In many respects, the evolution of gas turbine internal cooling technologies began in parallel with heat exchanger and fluid processing techniques, "simply" packaged into the constrained designs required of turbine airfoils; ie. aerodynamics, mechanical strength, vibrational response, etc. Turbine airfoils are after all merely highly specialized and complex heat exchangers that release the cold side fluid in a controlled fashion to maximize work extraction. Actively or passively cooled regions of the hot gas path in both aircraft engine and power generating gas turbines include the stationary vanes or nozzles, the rotating blades or buckets of the HPT stages, the shrouds bounding the rotating blades, and the combustor liners and flame holding segments. Collectively these components are referred to as the hot gas path (HGP). All such engines additionally cool the interfaces around the immediate HGP, thereby bringing into consideration the secondary flow circuits of the turbine wheelspaces and the outer casings that serve as both cooling and positive purge flows. The ever present constraints common to all components and systems include but are not limited to pressure losses, material temperatures, component stresses, geometry and volume, aerodynamics, fouling, and coolant conditions.

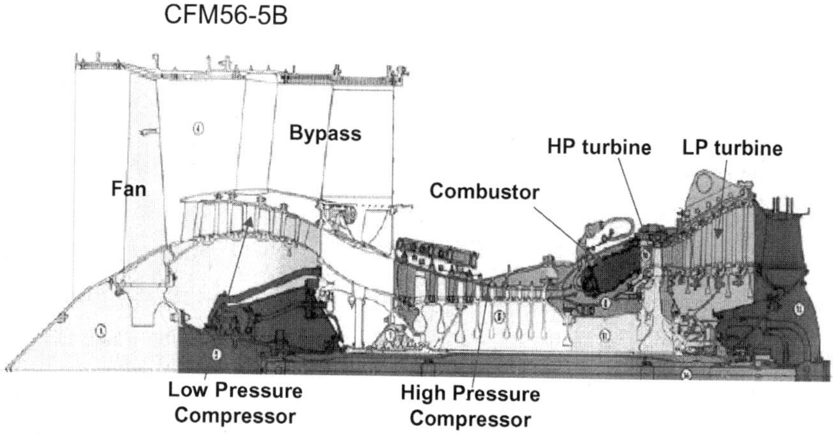

CFM56-5B

Figure 1. High bypass turbofan gas turbine engine

Cooling technology, as applied to gas turbine components such as the high-pressure turbine vanes and blades, is composed of five main elements that must work in harmony, (1) internal convective cooling, (2) external surface film cooling, (3) materials selection, (4) thermal-mechanical design, and (5) selection and/or pre-treatment of the coolant fluid. The enhancement of internal convective flow surfaces for the augmentation of heat transfer was initiated through the introduction of turbulators and pin-banks within investment cast airfoils. These surface enhancement methods continue to play a large role in today's turbine cooling designs. With the advancements in materials and manufacturing technologies of the last decade, a drastically larger realm of surface enhancement techniques has become cost effective for use in the cooling of turbine airfoils. Thc art and sciencc of film cooling concerns the bleeding of internal component cooling air through the external walls to form a protective layer of cooling between the hot gases and the component external surfaces. The application of effective film cooling techniques provides the first and best line of defense for hot gas path surfaces against the onslaught of extreme heat fluxes, serving to directly reduce the incident convective heat flux on the surface. Materials most commonly employed in cooled parts include high-temperature, high-strength nickel- or cobalt-based superalloys coated with yttria-stabilized zirconia oxide ceramics (thermal barrier coatings, TBC). The protective ceramic coatings are today actively used to enhance the cooling capability of the internal convection mechanisms and to dampen thermal gradients during transient events. The thermal-mechanical design of the components must marry these first three elements into a package that has acceptable thermal stresses, coating strains, oxidation limits, creep-rupture properties, and aero-mechanical response. Under the majority of practical system constraints, this means the highest achievable internal convective heat transfer coefficients with the lowest achievable friction coefficient or pressure loss. More often than not however, the challenge lies in the manufacturing limitations for such thermal-mechanical designs to be cost effective for the life of the components. The last cooling design element concerns the correct selection of the cooling fluid to perform the required function with the least impact on the cycle efficiency. This usually is achieved through the use of compressor air bled from the most advantageous stage of the compressor, but can also be done using off-board cooling sources such as closed-circuit steam or air, as well as intra-cycle and inter-cycle heat exchangers.

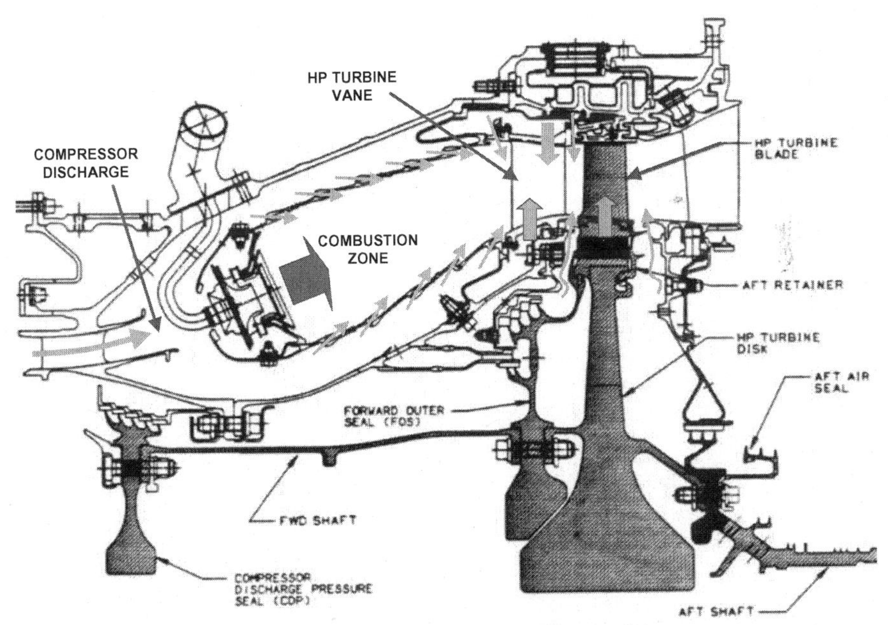

Figure 2. Gas turbine hot gas flow path

On a final introductory note, none of today's complex cooled turbine components would be possible without continuous advances in both materials and manufacturing science and technologies. Historically, the gains in HPT performance can be attributed two-thirds to cooling advances and one-third to materials advances; in some cases as much as 50/50. The turbine HGP is however also the most costly portion of the engine in terms of capital expenditure, operation, and maintenance/repair. As a consequence, improvements in materials and manufacturing, especially those that enable better cooling, carry a high return on investment. There are differing customer and market driven demands for aviation and power turbine engines, but always the need to enable turbine cooling through materials and manufacturing is key.

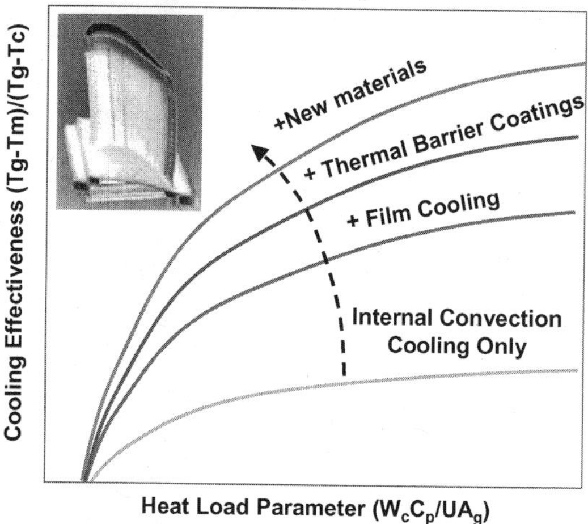

Figure 3. Cooling technology curves

RELATIONSHIP OF COOLING TO MATERIALS AND MANUFACTURING

The ultimate goals in gas turbine thermal management are consistent with the thermodynamics of isentropic energy transfer. By this is meant that (1) the component or some portion of it is desired to be isothermal to eliminate all internal thermal stresses, (2) the component is desired to be at or close to the hot gas temperature to minimize heat loads and inefficient energy transfers, and (3) thermal gradients from the component to the support and containment structure are desired to be minimized for these same reasons[2]. To the latter point, if the component and support structure can be thermally disconnected, then the desired spatial thermal gradient will be large while the separate material thermal gradients are small (i.e. insulated for reduced heat loss). This is essentially where gas turbine design began, and it is still the objective, only at a far higher operating temperature today. The advancement of turbine cooling has allowed engine design to exceed normal material temperature limits, but it has introduced complexities that have accentuated the thermal issues greatly. Cooled component design has consistently trended in the direction of higher heat loads, higher through-wall thermal gradients, and higher in-plane thermal gradients.

An overview of the HPT design system or design cycle is presented in the generic diagram of Figure 4. Cooling design analysis must include surrounding effects and constraints from aerodynamics, material properties, mechanical loads, lifing limitations, clearances etc. Analyses often must be combined thermal-mechanical predictions using very detailed finite element models, at times even sub-models of certain component sections. Most analyses are performed at one steady-state operating condition, e.g. 100% load, but more detailed analysis brings in the operational transient aspects to determine if requirements or constraints are violated under conditions such as normal start-up, fast start-up, takeoff, thrust-reverse, trips, and hot restarts. In all cases, engine experience design factors and known engine degradation factors must also be included. As examples, such factors may include the use of -3σ material properties, knockdown factors on cooling augmentation, and loss of coatings or metal thickness. A more specific look at the cooling design details is shown in Figure 5, including the various boundary conditions that may be affected by associated constraints, or that may affect the resulting critical lifing parameters. From these thermal boundary conditions can be seen what "material" matters most to the cooling of the turbine. These direct interfaces to cooling include material thermal radiation properties, geometry/shape and surface topology as affecting hot-side and cold-side heat transfer coefficients, substrate and coating thermal conductivity, specific heat, density, thermal diffusivity and thickness, temperature differences/gradients, wall temperature level, and cooling passage / hole flow coefficients. As materials and manufacturing can positively impact these factors, while also meeting the many other constraints, turbine cooling will advance further and faster.

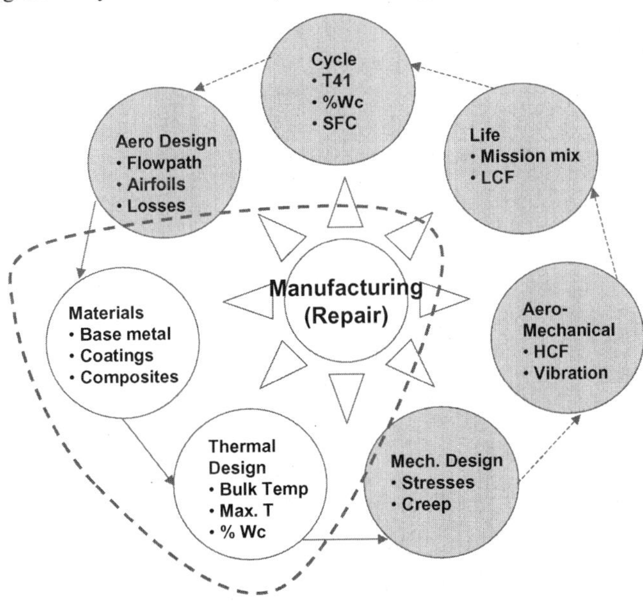

Figure 4. Turbine design cycle

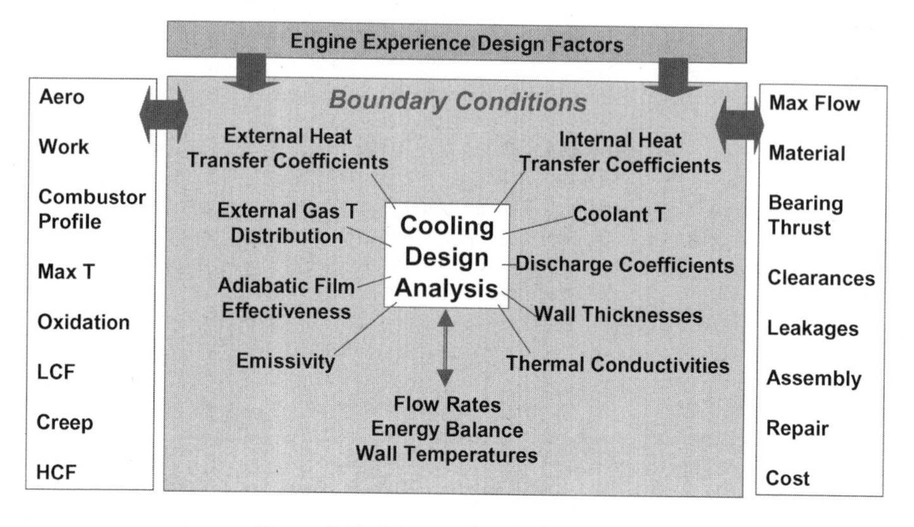

Figure 5. Turbine cooling design analysis

MATERIALS FOR (OR AS) TURBINE COOLING

Material categories within the turbine HGP include super alloys of various metallurgical composition, TBC's, MCrAlY bondcoats, oxidation and hot corrosion coatings, monolithic ceramics, ceramic composites, ceramic cores, and braze/joining compounds. With respect to turbine cooling, materials can be considered as beneficial cooling technologies in and of themselves through several mechanisms:

- Enable thermal properties while also having strength, oxidation resistance, foreign object damage resistance, etc.
- Enable machining and repair operations to achieve original net geometry
- Enable manufacturing processes
- Enable component architectures for cooling
- Enable heat flux rejection / shielding
- Enable complex ceramic core strength for casting

In the turbine airfoil design community and amongst the various manufacturers there are a very large number of cooled airfoil designs and an even larger number of individual cooling features. Figure 6a depicts a typical turbine blade having three distinct internal cooling circuits, similar to the actual blade shown in Figure 6b. The airfoil forward region is cooled by a radial passage that delivers impingement air to the leading edge through crossover holes, while the lead edge discharges showerhead film cooling. The airfoil mid-chord region is cooled by a five-pass serpentine with turbulated channels and 180-degree turns. The airfoil trailing edge region utilizes a radial passage with pin bank that feeds a distribution of small axial flow channels ending in bleed slots. The blade tip section has passage dust holes and dedicated coolant holes exiting into a tip cavity. In addition, rows of film holes other than the showerhead may be located to draw from any or all of the passages to refresh the film cooling. The typical cooled blade wall section is simply composed of an investment cast nickel or cobalt alloy base metal, a metallic bond coat, and a layer of TBC. The base metal structural design, thickness, and crystallographic orientation (equiaxed, directionally solidified, or single crystal) are an integral part of the overall thermal-mechanical design. An increase in base metal temperature capability serves to directly reduce cooling requirements, but usually also significantly increases cost. The base metal thickness provided as a product of the investment casting process will have a nominal value of 2 mm, with limits set on the thermal stresses, low cycle fatigue life, and creep rupture life. The bond coat is only about 0.2 mm thick. The bond coat is present to provide strain matching and bonding augmentation for the TBC, as well as some oxidation resistance. The TBC layer has come to be relied upon as a heat load mitigation technology. With a nominal thickness of 0.5 mm, TBC thickness variations can have a substantial effect on the external heat flux and the resulting maximum base metal and bond coat temperatures.

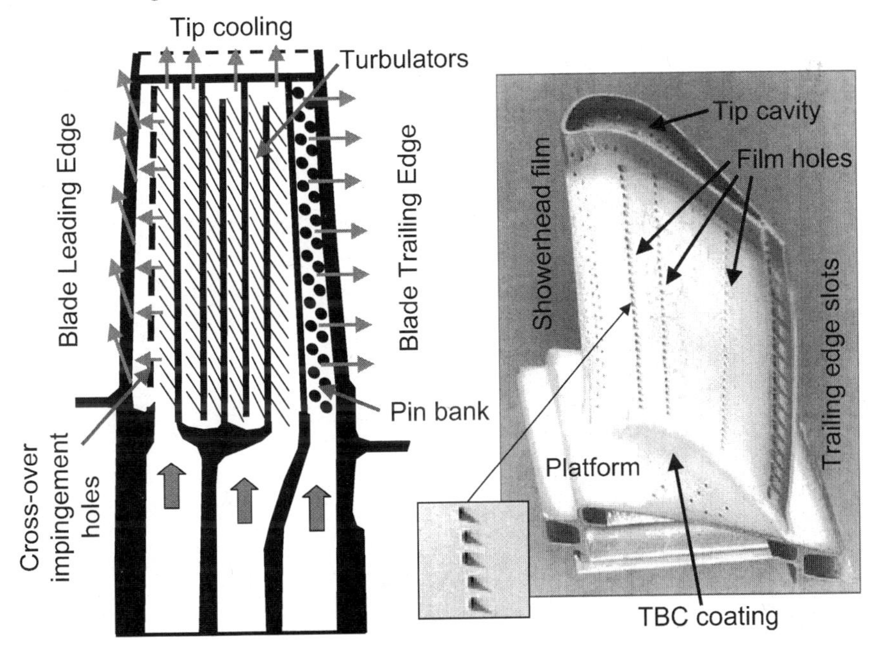

Figure 6. Highly cooled turbine blade

Reducing Incident Heat Flux via Materials

The incident heat flux on the HGP surfaces can be described after a manner in terms of three components, that due to basic aerodynamics, that due to surface condition, and that due to thermal radiation. The first of these components concerns design requirements that drive the pressure ratio and severity of convective heat transfer. Radiation is generally 5-10% of the heat flux, except in regions viewing or containing the flame, and while important is not a major factor except in the combustor. Surface condition may be primarily associated with roughness due to deposits, coating erosion, coating spallation, hot corrosion, and oxidation. Figure 7 shows a typical rough turbine blade from aircraft engine service, as well as surface topology scans of an applied TBC (unpolished and polished)[3]. An airfoil's aerodynamic surface roughness concerns both the initial manufactured roughness as the airfoil is first put into service, and also the in-service operational condition. The difference in roughness generated relatively quickly during operation and can be a factor of as much as four. Aerodynamic surface roughness is known to have large detrimental effects on heat load and drag, as well as emissivity, and so is a major factor requiring alleviation. Operational evidence suggests that an airfoil with initially smoother surface will not become as rough in service. Most high-pressure turbine blades now employ thermal barrier coatings (TBC), which are typically polished from a very rough condition (eg. for air plasma sprayed TBC) to a smoother condition. A typical specification for nominal initial average roughness is 2.5 microns.

Figure 7. Roughened turbine blade and TBC topography[2]
(reproduced by permission of ASME)

Certainly the most effective means for managing the cooling of the hot gas path components is to shield the surfaces from the incident heat flux. Before the introduction of ceramic coatings, film cooling was the primary method used to reduce the effective gas temperature driving heat flux to the surfaces. As long as a heat flux still exists, TBC can be effectively used to raise the external surface temperature, thereby also reducing heat flux to the surfaces. The usual design practice is to apply TBC while also decreasing the amount of coolant, which serves to improve efficiency, but tends to lead back to the original thermal gradients and heat flux levels. Aside from further increases in film cooling effectiveness for a given cooling flow rate, or the application of thicker TBC coatings, new methods are required to fundamentally reduce incident heat flux.

One potential method for the reduction of heat flux is to maintain a smooth surface condition throughout the service interval of the turbine. The lower temperature blade rows of compressors currently use classes of Teflon®-like coatings to produce an "anti-stick" surface that rejects deposits. Can analogous coatings be developed for the more aggressive high-temperature environments of turbines? If so, the coatings must be durable and also compatible with the TBC, without the possibility of infiltrating the TBC and causing early failure. It may be possible that alternate forms of TBC coatings could provide this anti-stick function while still serving as an insulating layer. Air plasma sprayed (APS) TBC is applied as particles of TBC quenched on the surface leading to a randomly distributed, somewhat layered microstructure as shown in Figure 8a[4]. Physical vapor deposited (PVD) TBC is a vacuum vapor process at the molecular level that results in a typical columnar microstructure as seen in Figure 8b[4]. Both APS and PVD TBC have porosity that is desirable for low thermal conductivity. If at least the exposed surface could be sealed and made smooth, then the airfoil would be less likely to collect deposits. An example of such an anti-stick turbine coating is described in patent literature for the prevention of build-up of corrosion products and other products of combustion from the hot gases. A tightly adherent anti-stick coating is applied over the environmentally resistance coating of the superalloy (eg. platinum aluminide or TBC), in which the aniti-stick coating is a metallic oxide consisting of a metal from the group of Pt, W, Si, Ge, Sn, and the group 5b elements of the periodic table[6]. Such specialty high temperature coatings may also provide modified radiative properties of benefit.

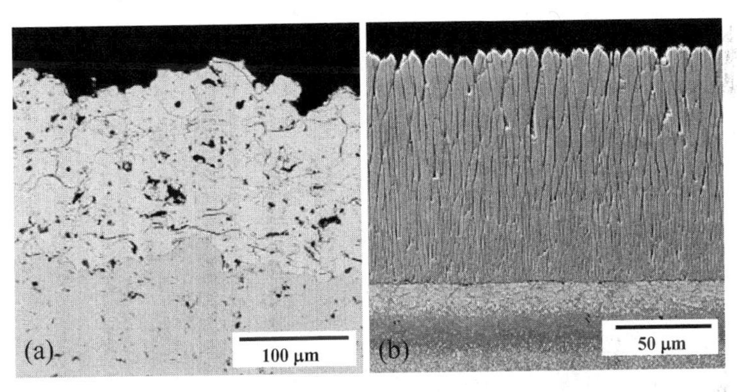

Figure 8. As-deposited YSZ (7 wt.% Y_2O_3): (a) APS; (b) EB-PVD.
Source: JI Eldridge, NASA Glenn Research Center, Cleveland, Ohio

Higher Temperature Materials

Hot gas path component thermal-mechanical design is a constantly evolving field that seeks to satisfy all the required strength and durability aspects of turbine service while also accommodating the desired aerodynamics and cooling. Though improved cooling technologies can allow greater flexibility in design, it is also true that designs producing lower thermal stresses will reduce the cooling flow burden, i.e. are optimized. This assumes that the design still meets an upper temperature material limit, or is not simply cooled to extreme levels. The impact of thermal stresses is lessened as the material temperature levels decrease, both in specific locations and in bulk, due to better material properties at lower temperatures. Therefore, the conventional approach to reducing thermal stress is simply to cool the region or component better. In most designs, this higher degree of local cooling leads to overall inefficiency and waste of cooling fluid. Higher strength materials can alleviate this situation, but new superalloys are very expensive and require many years to develop. New classes of materials with higher temperature capabilities, such as fiber reinforced ceramic matrix and metal matrix composites, are viewed to some day replace certain metal components in the turbine. These new materials however tend also to have much lower strength than metals.

Current research into advanced materials is heavily focused on the development of structurally capable high temperature monolithic ceramics and ceramic-matrix composites. In some applications, such materials can replace the use of TBC-coated metal, while in many HPT applications the ceramics still require cooling and alternate forms of TBC or environmental coating. Reduced cooling may be obtained through the increased allowable bulk material temperatures, but often the limiting factor is still the coating temperature limit and/or strain. An example is the development of Melt Infiltrated (MI) Ceramic Matrix Composites (CMC) for use in gas turbine engine applications. MI-CMC's consist primarily of continuous SiC reinforcing fibers within a matrix of SiC–Si, which is made using a molten Silicon infiltration process. The desirable properties of MI-CMC's include high thermal conductivity, high matrix cracking stress, high inter-laminar strengths, and good environmental stability (ie. oxidation resistance). In the pursuit of higher temperature capability combustors and turbines with maintained durability requirements, MI-CMC's are particularly suitable for gas turbine engine applications. Figure 9 shows the result of cyclic combustor rig testing comparing a stainless steel with a MI-CMC combustor liner, the latter exhibiting no evidence of distress[6]. Economics is still a limiting factor to the wide deployment of CMC's and similar materials systems.

Figure 9. Ceramic matrix composite combustor liner

MANUFACTURING TO ENABLE TURBINE COOLING

As the turbine design cycle of Figure 4 indicates, manufacturing technology is at the center of the multi-disciplinary approach, acting as key enabler to realize the final product, and to an ever-increasing extent also to realize component properties. Manufacturing is the connection between materials and design for the many critical aspects, including:

- Enable precision in geometry and features
- Enable flexibility in design, or expanded design space
- Enable cost reductions
- Enable structure / microstructure / architecture
- Enable lower variability (standard deviations)
- Enable surface topographies
- Enable component inspections
- Enable health monitoring and prognosis

Typical manufacturing operations employed in turbine HGP components include:

- Investment die and mold machining
- Investment casting
- Ceramic core fabrication
- Core leaching / abrasive cleaning
- CNC machining
- Electro-discharge machining (EDM)
- Electro-chemical machining (ECM)
- Shaped Tube ECM (STEM)
- Laser hole drilling
- Water jet drilling and machining
- Multiple braze operations
- Welding (TIG, electron-beam, laser)
- Metallic and ceramic coatings (vapor, plasma, slurry)
- Heat treatments

Film Cooling

Film cooling in all of its various formats has become a mainstay of turbine cooling technology today. To fully characterize film cooling behavior one would need a multitude of parameters concerning the film injection, hot gas flow, geometry, and interaction effects. Manufacturing constraints influence and limit the geometry of the film holes and the part. These factors include the effective film hole diameter, film hole length-to-diameter ratio, film hole axis angle to the external surface tangent, film hole orientation to the external and internal flow, film hole pitch-to-diameter ratio, and the specification of the hole exit shaping. All of these factors primarily affect the adiabatic film cooling, or do so indirectly by affecting the discharge coefficients. The effective film hole diameter refers to the combination of a measurable hole throat area and a discharge coefficient. This combination varies with the manufacturing method used to drill film holes[8]. Electro-discharge machining (EDM) uses a high electrical conductivity shaped tool to burn away the desired material within an electrolyte bath by electrical discharge. Burn rates are somewhat slow to avoid damage of the part, and to maintain good consistency of the resulting film hole. A diffuser shaped film hole, such as those shown in the inset of Figure 6b, is made with a negatively shaped tool. Film holes produced by EDM can usually be counted on to have very consistent discharge coefficients. Abrasive water jet machining is a relatively new process gaining popularity in small and micro drilling operations. Water-jet drilling can be applied by continuous or pulsed operation, percussion or trepanning operation. Water jet can be employed to produce virtually any of the film hole sizes and shapes in use today. Under controlled conditions, water jet can produce very clean and tailored holes, including shaped exits. Laser drilling is perhaps the most common technique employed due to its rapid processing, ready programming on multi-axis machines, and low cost. Laser drilling can also be applied by continuous or pulsed operation, percussion or trepanning operation. Film holes are usually drilled by the percussion method. In this method, a molten metal zone is formed as the laser energy is deposited in the metal, the metal is carried off as a vapor and also forms a plasma in the trapped region. As a result, laser drilling results in a somewhat irregular internal hole diameter and finish. For smaller holes this effect can be magnified to the extent that a significant percentage of the flow area cannot be measured by a simple pin gage.

Several alternate geometries of film cooling holes have been proposed within the last few years that have in some form demonstrated at least equivalent film effectiveness performance to the diffuser shaped holes noted above. These differing film holes may have specific form and function, either of limited or widespread potential, but each must also ultimately face the challenges of manufacturing, operability, and cost effectiveness. Emerging manufacturing methods to address more complex or intricate film hole shaping include micro-EDM (a.k.a. micro erosion), milling lasers, and pulsed electron beam. Micro EDM utilizes solid electrodes or wire electrodes with multi-axis control and pulsed power to erode fine details in geometry. Hole sizes down to 2-micron diameter with lengths of up to ten times the diameter are possible. At such sizes, grinding and polishing operations are also feasible around the film holes. Unlike a drilling laser, a milling laser uses gigawatt level excitation on the order of nanoseconds to directly vaporize metal rather than melting it. This leads to sculpting and three-dimensional milling capability at a 10 kHz rate with a laser diameter of about 10 microns. Compared to a drilling laser though, the milling laser process is very slow, but might be used for detail shaping. Pulsed electron beam is typically used for rapid drilling of fine mesh screens and plates. This process can drill very repeatable and clean holes down to a diameter of 100 microns. Pulsed electron beam can be used with any material though its limitation is depth.

Micro Cooling[2]

All cooled turbine airfoils in commercial operation today utilize cooling channels and film holes that are considered to be macroscopic in physical magnitude. The generally accepted delimiters between macro and micro cooling are (1) whether the feature (internal passage) can be repeatedly manufactured via investment casting methods, and (2) whether the flow passage (film hole) may become plugged by particles in the cooling fluid. The acceptable sizes are relative to the turbine size and the operating environment. Aircraft engine turbine airfoils are small and require smaller flow passages and film holes, while large power turbine castings cannot support equivalent small passages and would more easily be plugged if the same size film holes were employed. As an example, an average aircraft engine may allow film hole diameters as small as 0.457 mm (.018 inches), while an average heavy frame power turbine may be limited to a minimum diameter of only 1.02 mm (.040 inches).

The concept of micro cooling for airfoils is the natural application of thermodynamics and heat transfer to accomplish two main goals. First, to spread out the cooling network in a series of smaller and highly distributed channels, or sub-channels, providing better uniformity of cooling and lower magnitude in-plane thermal gradients. This is analogous to the distribution of blood vessels in the human body. Second, to bring the cooling fluid closer to the outer surface of the airfoil, thereby creating a more efficient transfer of heat, one that involves less of the bulk component material. Micro cooling may be thought of as a complete airfoil cooling solution, or as a regional cooling device (e.g. leading edges). Figure 10 provides a schematic version of generic micro cooling for an airfoil using one form of a transpiration surface.

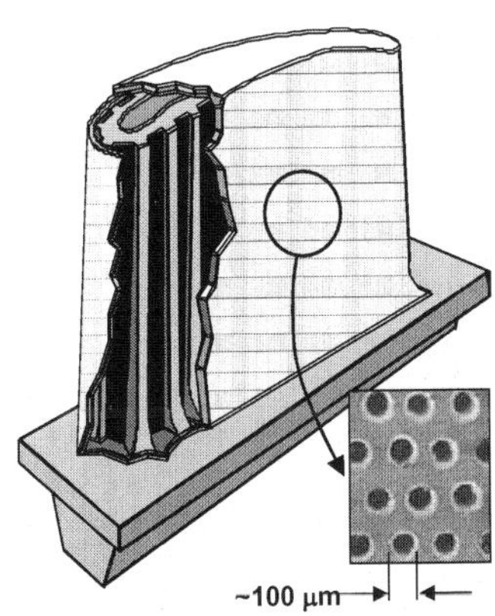

Figure 10. Micro cooled turbine airfoil with electron beam holes[2]
(reproduced by permission of ASME)

The most notable developments of micro cooling are those of the Allison Advanced Development Corporation[9], known by the trade names of Lamilloy® and CastCool®. The airfoil walls in Lamilloy are formed as two or more sheets of metal bonded together with distributed pins, while those of CastCool are investment cast and limited to a double-wall geometry. The cooling flow is introduced via impingement jets that are staggered with respect to the exiting film holes. Full cooling distribution is obtained, and a form of transpiration-like film cooling results from the surface normal oriented holes in the outer layer. The challenges of this micro cooling include hole plugging, wall strength, film cooling, manufacturing, and cost. Russian and Ukrainian turbine airfoil research developed a shell-and-spar approach to manufacturing[10]. In this method, the interior portion of the airfoil is cast in a simple format, such as the spar portion in Figure 10, and cooling channels machined in the exposed surface. The channels are then filled with a leachable material, a thin outer airfoil skin is bonded to the non-channel metal regions, and the filler is leached out. The result is a distribution of small cooling channels close to the outer surface. Again, these micro cooling forms face significant material and processing challenges to meet the durability and cost requirements for gas turbines.

Hot Gas Path Surface Contouring[2]

The aerodynamic design of the hot gas path emphasizes smoothly varying surfaces, preferably with small airfoil leading edge diameters and razor thin trailing edges. Airfoil lean, bow, and twist are now more commonplace to improve efficiency and reduce hot gas secondary flow losses. Bounding endwalls at the hub and casing are generally circular sections at any particular axial location through the turbine. The locations at the hub and casing where the airfoils meet the endwalls usually only have a small manufacturing fillet as the surface transition. These locations are the sites of the remaining major secondary flow losses generated in the turbine.

Recent research has begun to investigate the use of macro level contoured flow path endwalls, or non-axisymmetric endwalls, as well as alternate shaping of airfoil-endwall fillet regions. An example of complex endwall contouring is shown in Figure 11. Contoured endwalls may lead to significant aerodynamic efficiency gains and improved cooling. A general rule of HGP loading is that local heat loads and aero efficiency vary in proportion to the secondary flows predominantly associated with the endwall regions of the hub and casing. Since a very large portion of the total HGP is influenced by these secondary flows, the resulting sensitivity of the overall heat load and the local part life can be a strong function of the secondary flows. Present studies have demonstrated gains of around +0.5% in HPT aerodynamic though non-axisymmetric contouring of the vane and blade endwalls. Projection of such gains through the fan, compressor, and turbine stages, leads to potential aerodynamic efficiency increases of several points. Combined with consequent advantages in cooling flow reductions due to better aerodynamics, the overall entitlement for engine cycle efficiency could be as much as +2%. Micro level surface contouring, one or two orders of magnitudes less than the overall airfoil size, also has the potential to influence local aerodynamic and cooling phenomena in the hot gas path. Surface riblets aligned in the streamwise direction, similar to those tested on aircraft fuselages and ship hulls, may reduce both drag and heat transfer coefficients by 5 to 10%. Can such surface contours be economically manufactured though? An emerging manufacturing technology that may enable this is direct metal laser sintering (DMLS)[11], also called laser net shape manufacturing. In this process, a controlled laser is used to melt powder metal of the desired composition for consolidation as a solid phase. The challenges here lie in the resulting ability to create crystallographic orientation and hence the required material properties.

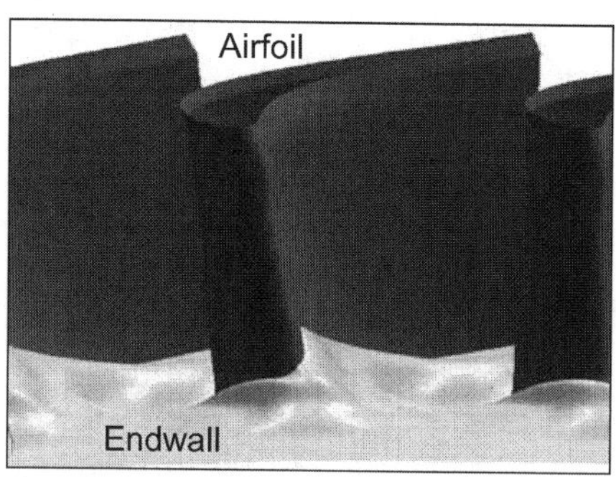

Figure 11. Turbine blade with complex endwall contouring[2]
(reproduced by permission of ASME)

Surface Heat Transfer Augmentation[12]

In addition to pure materials solutions, thermal-mechanical design can use cooling methods to reduce thermal stresses. As previously noted, greater degrees of cooling uniformity, or greater local control over specific cooling magnitudes, can be utilized to alter local thermal gradients. Several forms of patterned surface augmentation, amenable to investment casting, constitute the bulk of conventional internal cooling techniques. These include pin or pedestal arrays and regularly spaced turbulators or rib rougheners. The intent of these surface augmentation methods is to derive a greater benefit than simple wetted surface area increase by also creating additional vortices and turbulence. The use of impingement jets for the cooling of various regions of modern gas turbine engines is also widespread, most especially within the high-pressure turbine. Since the cooling effectiveness of impingement jets is very high, this method of cooling provides an efficient means of component heat load management. Target surface roughness and texturing can have a major impact on the impingement dominated heat transfer portion of the surface. Moreover, discrete roughness can frequently be tailored to address cooling needs in specific localities. Effects that tend to thin the boundary layer relative to the roughness element heights, such as increased Reynolds number or decreased jet diameter, lead to increased heat transfer. Heat transfer coefficients have been observed to increase by 25 to 50% due to the use of a rough surface whose increased wetted area was only 10%. Moreover, the ratio of peak-to-average heat transfer can be reduced by as much as half. Another class of surface enhancements results from the depression of features into the cooling channel or surface walls, forming recesses rather than projections. Generically, such features are known as concavities (or dimples), and may be formed in an infinite variation of geometries with various resulting heat transfer and friction characteristics. In the broadest sense, these concavity surface flows are one of a larger category known as "vortex" technologies, which include various means for the formation of organized vortical or swirling flows in turbines.

The limitations of investment casting may be overcome by manufacturing technologies such as electron beam sculpting, particle loaded braze tapes or slurries, pulsed ECM (PECM), and here too DMLS. The Surfi-Sculpt® technology[13] uses an electron beam to locally make surface material molten to the degree that it may be maneuvered into a wide range of patterns and textures. Scanning electron images of surfaces produced by this method[13] indicate the ability to form highly repeatable features of complex geometry with projecting elements about 3 mm tall, or honeycomb type structures (or projected cell walls) with 3 mm cell widths. The possible geometries are nearly limitless, as long as material surface damage and engineering flaws are also constrained. Brazing technology can be applied to more than component joining as shown in Figure 12. In this example, a high temperature nickel alloy braze tape has been loaded with nickel particles and applied to both the interior and exterior of a Hast-X® tube[14], Figure 12a. The carrier adhesive that temporarily bonds the so-called micro turbulators to the surface is burned away during brazing, leaving the very clean and sealed augmented surface of Figure 12b. While ECM processes typically remove material only very near the tip of the tool, a recently developed PECM process can be applied to remove patterns of material along the entire length of the tool[15]. This process starts with a polymer coated ECM delivery tube/electrode, removes the coating (eg. by laser ablation) in a predefined pattern to expose the electrode, and uses pulsed current through the entire electrode to yield a complete surface pattern in the finished component.

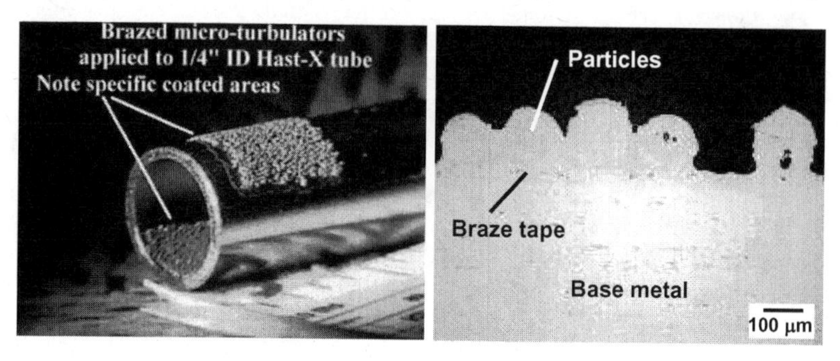

Figure 12. Brazed micro turbulators

Sensor Technologies

Finally, although most of the emphasis in materials and manufacturing for gas turbines concerns improved engine performance, sensing technologies play a key role in engine operation as well as conditioned-based maintenance. Increasingly, materials and manufacturing can now enable improvements in sensing, including in-situ instrumentation. Direct-Write is a process currently used in manufacturing passive and active electronic components and interconnects in printed circuit boards[16]. A desired metal or ceramic powder is formulated as an ink slurry for computer controlled, maskless surface deposition at room temperature in any pattern. Heat treatment provides consolidation of the pattern. The simultaneous pattern and material transfer over any surface can be achieved with extreme precision, resulting in a functional structure or a working device/instrument. Figure 13 shows an example of a thin film thermocouple deposited by the Direct-Write process on a turbine blade, where the thermocouple junction is formed on the blade metal and the leads are deposited over a dielectric layer of ceramic.

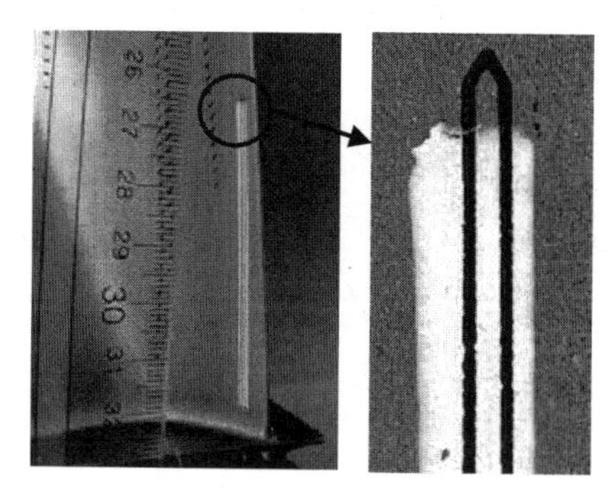

Figure 13. Thermocouple applied by Direct Write

CONCLUSION

Gas turbines contribute a significant and growing fraction of the world's electric energy demand, as well as the majority of aerospace propulsion. The turbine HGP is the most costly portion of the engine and as a consequence, improvements in materials and manufacturing, especially those that enable better cooling, carry a high return on investment. The need to enable turbine cooling through materials and manufacturing technology advances is key. This summary has identified many of the elements in turbine cooling that can benefit from enabling materials and manufacturing. Several emerging technologies have been presented with the capability to significantly enhance turbine performance and/or durability. On a final note, it should be remembered that many of the technologies noted, though developed with new parts in mind, may have equal or greater benefit when applied to repair and reconditioning operations.

REFERENCES

[1]R.S. Bunker, Turbine Cooling Design Analysis, in *The Gas Turbine Handbook*, U.S. Dept. of Energy, Office of Fossil Energy, National Energy Technology Lab, DOE/NETL-2006/1230 (2006).

[2]R.S. Bunker, Gas Turbine Heat Transfer: 10 Remaining Hot Gas Path Challenges, *Journal of Turbomachinery*, **129**, 193-201 (2006).

[3]R.S. Bunker, R.S., The Effect of Thermal Barrier Coating Roughness Magnitude on Heat Transfer With and Without Flowpath Surface Steps, Paper No. 41073, International Mechanical Engineering Conference, Washington DC (2003).

[4]J.I. Eldridge, NASA Glenn Research Center, Cleveland, Ohio.

[5]W.R. Stowell, B.A. Nagaraj, C.P. Lee, J.F. Ackerman, and R.S. Israel, Enhanced Coating System for Turbine Airfoil Applications, US Patent 6,394,755 (2002).

[6]G.S. Corman, A.J. Dean, S. Brabetz, M.K. Brun, K.L. Luthra, L. Tognarelli, and M. Pecchioli, Rig and Engine Testing of Melt Infiltrated Ceramic Composites for Combustor and Shroud Applications, Paper No. 2000-GT-638, International Gas Turbine Conference, Munich, Germany (2000).

[7]R. Bunker, T. Simon, D. Bogard, A. Schulz, A. Burdet, and S. Acharya, Film Cooling Science and Technology for Gas Turbines: State-of-the-Art Experimental and Computational Knowledge, *Von Karman Institute for Fluid Dynamics Lecture Series VKI-LS 2007-06*, ISBN-13 978-2-930389-76-1, Brussels, Belgium (2007).

[8]R.S. Bunker, The Effects of Manufacturing Tolerances on Gas Turbine Cooling, Paper No. GT2008-50124, International Gas Turbine Conference, Berlin, Germany (2008).

[9]P.C. Sweeney and J.P. Rhodes, An Infrared Technique for Evaluating Turbine Airfoil Cooling Designs, *Journal of Turbomachinery*, **122**, 170-177 (1999).

[10]A.S. Novikov, S.A. Meshkov, and G.V. Sabaev, Creation of High Efficiency Turbine Cooled Blades With Structural Electron Beam Coatings, *in collection of papers* Electron Beam and Gas-Thermal Coatings, Paton IEW, Kiev, 87-96 (1988).

[11]M.W. Khaing, J.Y.H. Fuh, and L. Lu, Direct Metal Laser Sintering for Rapid Tooling: Processing and Characterisation of EOS Parts, *Journal of Materials Processing Technology*, **113**, 269-272 (2001).

[12]R.S. Bunker, Innovative Gas Turbine Cooling Techniques, in *Thermal Engineering in Power Systems*, ISBN 978-1-84564-062-0, Wessex Institute Technology Press, Southampton, Great Britain (2008).
[13]B. Dance, A. Buxton, An Introduction to Surfi-Sculpt® - New Opportunities New Challenges, 7th International Conference on Beam Technology, Halle, Germany, April 17-19 (2007).
[14]W.C. Hasz, R.A. Johnson, C.P. Lee, M.G. Rettig, N. Abuaf, and J.H. Starkweather, Article Having Turbulation and Method of Providing Turbulation on an Article, U.S. Patent 6,598,781 (2003).
[15]R.S. Bunker and B. Wei, Cooling Passages and Methods of Fabrication, U.S. Patent 6,644,921 (2003).
[16]D. Mitchell, A. Kulkarni, E. Roesch, R. Subramanian, A. Burns, J. Brogan, R. Greenlaw, A. Lostetter, M. Schupbach, J. Fraley, and R. Waits, Development and F-Class Industrial Gas Turbine Engine Testing of Smart Components with Direct Write Embedded Sensors and High Temperature Wireless Telemetry, Paper No. 2008GT-51533, International Gas Turbine Conference, Berlin, Germany (2008).

Energy Materials

SYNTHESIS, SINTERING AND DIELECTRIC PROPERTIES OF NANO STRUCTURED HIGH PURITY TITANIUM DIOXIDE

Sheng Chao and Fatih Dogan
Missouri University of Science and Technology
Rolla, Missouri, USA

ABSTRACT

High purity nanosized titanium dioxide (TiO_2) powders were synthesized by precipitation method using Ti(IV)-isopropoxide as starting material. Well-crystallized and phase pure anatase TiO_2 powders with a particle size about 10nm can be obtained by calcination of freeze-dried precipitates at 400°C. The sinterability of powders calcined at 400°C and 700°C were compared with commercial TiO_2 powders with similar particle size. The electrical and dielectric properties of the TiO_2 bulk samples sintered at various temperatures were measured and correlated with the microstructural development.

INTRODUCTION

Due to its wide applications such as photocatalysts,[1] solar cells,[2] gas sensors[3] and dielectric materials[4], TiO_2 has attracted great scientific and technological interest. TiO_2 powders are commonly prepared by means of the sulphate process, the chloride process,[5] or sol-gel process.[6] In this study, high purity and nanosized TiO_2 powders were prepared to obtain nanostructured dielectric materials with low loss factor and leakage current.

EXPERIMENTAL

High purity nanosized TiO_2 powders were prepared by precipitation of Ti(IV)-isopropoxide followed by freeze-drying of the precipitates. 50ml Ti(IV)-isopropoxide ($Ti[OCH(CH_3)_2]_4$ 99.995%, Alfa-Aesar) was added into 1L deionized water under constant stirring. The precipitate was mixed with 1L of 2-Propanol (99.5% Alfa-Aesar) and ultrasonicated (SONICS, Vibra-Cell, Newtown, CT, USA) for 1 minute to deflocculate the hydroxides. Gel-like precipitates were freeze-dried (Genesis SQ Freeze Dryers, Winchester, Hampshire,UK) at -25°C under vacuum for 72hrs. Fluffy precursor powders were calcined at various temperatures (400°C, 500°C, 600°C and 700°C) for 1h with a heating and cooling rate of 2°C/min and 5°C/min, respectively. The crystallinity and phase composition of the calcined powders were determined by X-ray diffraction (XRD, Philips X'Pert, Holland). The size and morphology of powder were observed by Scanning Electron Microscopy (SEM, Hitachi S-4700, Japan).

Pellets were prepared by uniaxial pressing at 50MPa followed by iso-static pressing at 300MPa using powders calcined at 400°C and 700°C. Commercially available TiO_2 powders (99.9%, Sigma Aldrich) were also pressed into pellets following the same procedure. Sintering was performed at 900°C, 950°C and 1000°C for 2hrs in oxygen atmosphere with a heating rate of 4°C/min.

Samples about 10mm in diameter and 0.8mm in thickness were used for electrical property measurements. Silver paste was painted on the samples as top and bottom electrodes and cured at 500°C. D.C. conductivity measurements were conducted using an electrometer (Model 6517 Keithley Instruments) in ambient atmosphere. Relative dielectric constant was calculated according to capacitance measured on a Solartron 1260 impedance analyzer connected with a Solartron 1296 dielectric interface (Solartron analytical) in a frequency range of 1Hz to 1MHz, with a voltage amplitude of 1V.

RESULTS AND DISSCUSION

Hydrolysis of the Ti(IV)-isopropoxide resulted in formation of titanium hydroxide precipitates. XRD analysis (Fig. 1) reveals that there is no crystalline phase (absence of diffraction peaks) in as-prepared powders. Well-crystallized anatase TiO_2 were obtained by calcination at temperature as low as 400°C. Further increase the calcination temperature leads to sharper diffraction peaks, indicating higher degree of crystallinity and larger crystallite size. At 700°C, a small amount (~9%) of rutile phase was identified. This anatase to rutile phase transition temperature is lower than commonly observed temperature around 915°C,[7] which may be attributed to the difference in particle size of hydroxide precursors. The temperature of phase transformation was found to be lowered in nanosized powders.[8] The commercial powders were found to be a mix of anatase phase and rutile phase, with an average crystalline size close to that of the powders calcined at 700°C. The phase and the crystallite size of the powders determined by XRD are summarized in Table 1.

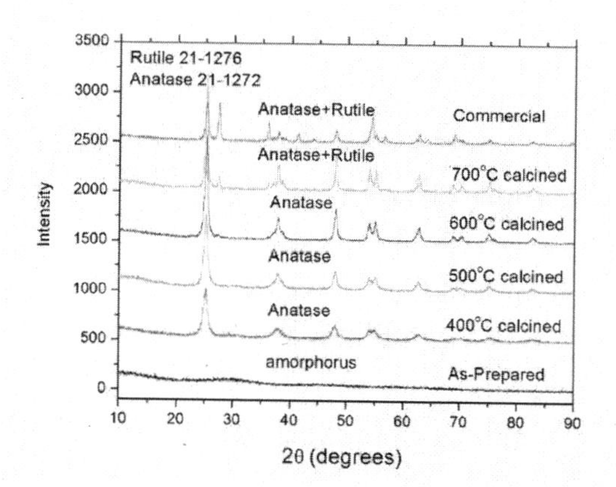

Figure 1. XRD pattern of the TiO_2 powders

Table 1. Crystalline phases and the crystallite size of TiO_2 powders determined by XRD

Calcination Temperature	**Phase**	**Crystalline size (nm)**
400°C	Anatase	9.7±3
500°C	Anatase	11.8±5
600°C	Anatase	15.7±4
700°C	Anatase:Rutile (91:9)	21.3±5
Commercial Powder	Anatase:Rutile (49:51)	17.5±5 (A), 26.9±9 (R)

The morphology of powder, observed by SEM, are shown in Figure 2. It is revealed that at calcination temperatures below 600°C, the average particle size of powders is similar, which is in good agreements with the XRD results. After calcination at 600°C, particles in size of about 20nm form a certain degree of agglomeration. It is apparent that some partial sintering (neck formation) occur between the particles for powders calcined at 700°C. It was noticed that the particle size and shape of powders are very uniform at all calcination temperatures. The powder morphology of the commercial powders looks similar to that of the powder calcined at 600°C while some larger agglomerates appear to be present in commercial powders.

Figure 2. SEM images of the TiO_2 powders (a) as-dried; (b) calcined at 400°C; (c) calcined at 500°C; (d) calcined at 600°C; (e) calcined at 700°C; (f) commercial powder (bar scale: 500nm)

The samples prepared by using powders (calcined at 400°C and 700°C as well as commercial powders) were sintered at 900°C, 950°C and 1000°C. The sintering of the samples was carried out in oxygen atmosphere to prevent the formation of oxygen vacancies which could significantly affect the dielectric properties of TiO_2. Figure 3 shows the relationship between relative density and sintering temperature of the samples. The sintering density of TiO_2 samples prepared in this work is significantly higher than that of the samples prepared using commercial powders. After sintering at 1000°C, the density of the samples from precipitated powders is near 98%, whereas the density of the samples from commercial powder is about 92%. This may be attributed to the difference in the crystalline phase of these powders as 400°C calcined powders are pure anatase; powders calcined at 700°C have 91% of anatase phase, while commercial powders have only 49% anatase phase. It was found that the densification process of TiO_2 can be assisted by the anatase to rutile phase transition.[9] Therefore, powders with anatase phase tends to densify at lower temperatures.

SEM images of the as-fired surfaces of the TiO_2 ceramics sintered at different temperatures are shown in Fig. 4-6. It was observed that as the sintering density gradually increases from 900°C to 1000°C, there is no significant grain growth within this temperature range. In addition, there is no obvious difference in grain size of the samples prepared from powders calcined at 400°C and 700°C. It should be mentioned that the radial shrinkage of the sample made of 400°C calcined powder is 8.5% larger than that of the sample prepared from 700°C calcined powders due to the differences in their green densities

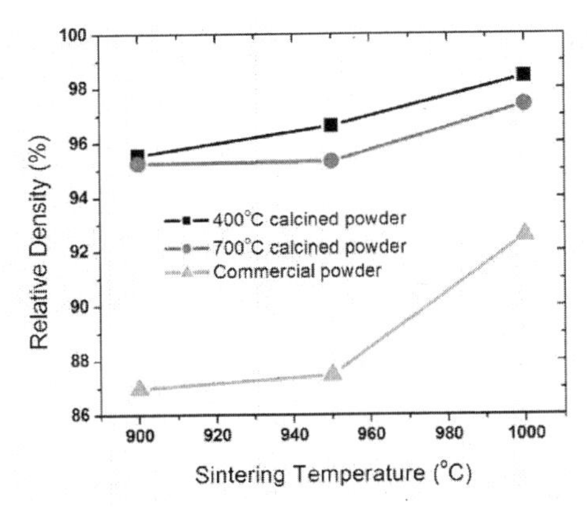

Figure 3. Relative density vs. sintering temperature of TiO_2 ceramics

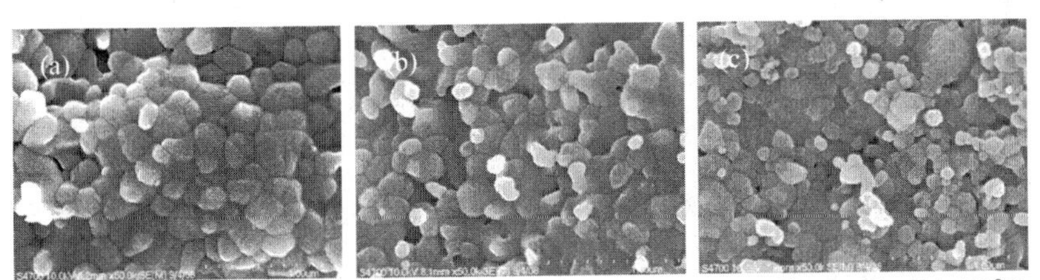

Figure 4. SEM images of the surface of TiO_2 ceramics sintered at 900°C for 2hrs (a) 400°C calcined powders; (b) 700°C calcined powders; (c) commercial powders (scale bar: 1μm)

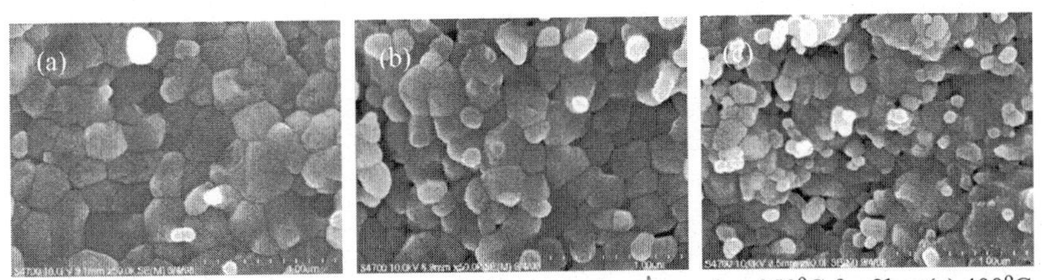

Figure 5. SEM images of the surfaces of TiO_2 ceramics sintered at 950°C for 2hrs (a) 400°C calcined powders; (b) 700°C calcined powders; (c) commercial powders (scale bar: 1μm)

Figure 6. SEM images of the surfaces of TiO_2 ceramics sintered at 1000°C for 2hrs (a) 400°C calcined powders; (b) 700°C calcined powders; (c) commercial powders (scale bar: 1μm)

Dielectric constant and loss tangent of sintered samples are shown in Figure 7. It was observed that samples sintered at higher temperature reveals higher dielectric constant, due to increased relative density. Commercial powder sintered at 900°C has high apparent dielectric constant and high loss tangent at low frequencies. This feature is characteristic for the conduction loss due to relatively high electric conductivity as a result of low sintering densities (~87%). Since open porosity can serve as fast conduction pathways,[10] porous TiO_2 ceramics are electrically more conductive than dense counterparts. Increasing the sintering density of TiO_2 made from commercial powders to about 93%, significantly reduced its dielectric loss from 0.249 to 0.0619 at 10Hz, due to reduced conductivity as shown in Table 2. As the relative density of TiO_2 samples prepared by using of calcined powder is much higher, improved dielectric properties were observed with respect to non-dispersive dielectric constant, low dielectric loss and low conductivity.

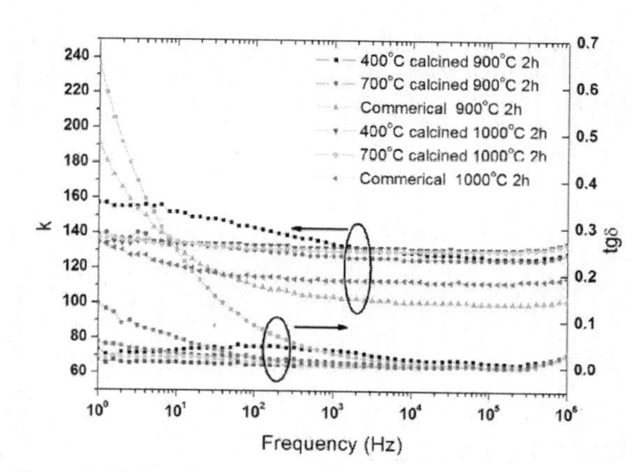

Figure 7. Dielectric constant and loss tangent of TiO_2 ceramics

Table 2. D.C. conductivity of TiO_2 ceramics

Powder Source	Calcination Temperature	Sintering Temp.	Conductivity (S/cm)
Titanium Isopropoxide	400°C	900°C 2h	1.94E-12
Titanium Isopropoxide	400°C	1000°C 2h	4.54E-13
Titanium Isopropoxide	700°C	900°C 2h	1.19E-12
Titanium Isopropoxide	700°C	1000°C 2h	6.27E-13
Commercial	---	900°C 2h	2.73E-11
Commercial	---	1000°C 2h	1.57E-12

CONCLUSIONS

High purity TiO_2 nanopowders were synthesized by precipitation method using Ti(IV)-isopropoxide as starting compound. Well-crystallized powders with a crystallite size of 10nm were obtained by calcination at temperatures as low as 400°C. The anatase to rutile phase transition temperature was lowered to about 700°C, due to very small particle size of powders. The sinterability of the powders with higher amount of anatase phase than rutile was improved in that the densification process may have been assisted by the anatase to rutile phase transition. Electrical conduction loss was found to be the dominant loss mechanism in TiO_2. Higher sintering densities lead to reduced conductivity, which, in turn, resulted in reduced dielectric loss.

ACKNOWLEDGEMENT

This work was funded by a MURI program sponsored by Office of Naval Research under Grant No. N000-14-05-1-0541.

REFERENCES

[1]Q. Shi, D. Yang, Z. Jiang and J. Li, Visible-light Photocatalytic Regeneration of NADH Using P-doped TiO2 Nanoparticles, *J. Mol. Cata. B: Enzymatic* **43**, 44–48 (2006).

[2]B. O. Regan and M. Graetzel, A Low-Cost, High-Efficiency Solar Cell Based on Dye-Sensitized Colloidal TiO_2 Films, *Nature (London)*, **353,** 737–40 (1991).

[3]Skubal, L. R., Meshkov, N. K. and Vogt, M. C., Detection and Identification of Gaseous Organics Using a TiO_2 Sensor. *J. Photochem. Photobiol. A: Chem.*, **148**, 103–108 (2002).

[4]C.T. Dervos, Ef. Thirios, J. Novacovich, P. Vassiliou and P. Skafidas, Permittivity Properties of Thermally Treated TiO_2, *Mater. Lett.*, **58**, 1502-1507 (2004).

[5]H. Jensen, A. Soloviev, Z. Li and E. G. Søgaard, XPS and FTIR Investigation of the Surface Properties of Different Prepared Titania Nano-powders, *Appl. Surf. Sci.*, **246**, 239–249 (2005).

[6]P.D. Moran, J.R. Bartlett, G.A. Bowmaker, J.L.Woolfrey and R.P. Cooney, Formation of TiO_2 Sols, Gels and Nanopowders from Hydrolysis of Ti(OiPr)4 in AOT Reverse Micelles, *J. Sol–Gel Sci. Technol.* **15**, 251–262 (1999).
[7]I.E. Campbell and E.M. Sherwood (Eds.), High-Temperature Materials and Technology, Wiley, New York, USA, 1967.
[8]Y. Hu, H. -L. Tsai, C.-L. Huang, Phase Transformation of Precipitated TiO_2 Nanoparticles, *Mater. Sci. Eng. A*, **344**[1-2],209-214 (2003).
[9]S.-C. Liao, Kook D. Pae and W. E. Mayo, High Pressure and Low Temperature Sintering of Bulk Nanocrystalline TiO_2, *Mater. Sci. Eng. A*,**204**[1-2], 152-159 (1995).
[10]S.-H. Song, X. Wang and P. Xiao, Effect of Microstructural Features on the Electrical Properties of TiO_2, *Mater. Sci. Eng. B*, **94**[1], 40-47 (2002).

SORPTION/DESORPTION PROPERTIES OF MgH_2-OXIDE COMPOSITE PREPARED BY ULTRA HIGH-ENERGY PLANETARY BALL MILLING

Y. Kodera[1], N. Yamasaki[1], J. Miki[1], M. Ohyanagi[1], S. Shiozaki[2], S. Fukui[2], J. Yin[3], T. Fukui[3]
[1]Department of Materials Chemistry, Innovative Materials and Processing Research Center, Ryukoku University, Otsu Shiga 520-2194, Japan
[2]Kurimoto, LTD
[3]Hosokawa Powder Technology Research Institute
Email: ohyanagi@rins.ryukoku.ac.jp

ABSTRACT

The metal hydride-oxide composite of MgH_2 with 50wt% Al_2O_3 was prepared by mechanical grinding (MG) process with ultra high-energy planetary ball mill for a short time. Conventionally, planetary ball mill, in which a belt transfers the driving power to the mill vial, had been used for 20 to 200 h in order to improve hydrogen sorption/desorption properties of MgH_2 system. However, ultra high-energy planetary ball mill with a gear wheel to the transfer was selected. The influence of gravitational acceleration (G) as the milling intensity during MG on those properties was significant. The MG process for 10 min with 150 G made MgH_2 to have the particle size of tens of nano-meters and homogeneous distribution with oxides. When the specimen was heated with 5 °C/min under 0.001 MPa of hydrogen, the desorption temperature of MgH_2 with 50wt% Al_2O_3 was 220°C.

INTRODUCTION

For future hydrogen technology, one of the major tasks is the development of an inexpensive, safe, and lightweight storage facility for mobile application. There are several methods of hydrogen storage. The use of hydrogen storage metals or compounds is being considered as the most promising method. To prepare those materials and improve their hydrogen storage properties (such as a hydrogen sorption/desorption kinetics and temperature), many processes have been developed. A high-energy ball milling using a planetary system is one of these processes. Generally, the planetary ball mill is used, for a short time, to prepare inorganic nano-powder, because the system can supply a higher input energy against unit time. In the case of hydrogen storage material, this system is often selected in order to improve the hydrogen storage properties.

Among other candidates, magnesium hydride (MgH_2) is a very attractive material, which exhibits a lightweight, low production cost, and high hydrogen capacity of 7.6wt%. However, MgH_2 has disadvantages such as slow hydrogen sorption/desorption (hydriding/dehydriding) kinetics, high hydriding/dehydriding temperatures, and its high reactivity to oxidation. Especially, the dehydriding reaction of MgH_2 is too slow to form the basis of a practical hydrogen store. Therefore, many researchers have focused on improving the hydriding/dehydriding kinetics of the Mg system. Significant progress has been made by using nanocrystalline MgH_2 powder produced by the high-energy ball milling, and by using additives of transition metals, or their oxides, during the milling process [1,2]. However, an extremely long milling time (20 to 200h and more) was often required to observe the drastic improvement of the hydriding/dehydriding kinetics of MgH_2 [1,2] This issue confronted not only the Mg-based system but other systems as well.

The study of the milling process is necessary in order to be able to obtain a system with efficiency, productivity, and the ability to prepare hydrogen storage materials with improved hydrogen storage properties. The extremely long milling time is a significant disadvantage even if a material with attractive hydrogen storage properties is produced. However, many researchers, who focused on improving properties, did not pay much attention to the milling process. The influence of planetary ball milling conditions (e.g. the revolution speed of a mill vial, the revolution radius of the mill vial, rotation-to-revolution speed ratio, ball filling fraction in the mill vial, ball-to-powder weight ratio, milling time) on the sample properties is extremely complicated. Moreover, the milling conditions have limitations depending on the equipment design. Conventionally, a planetary ball mill, in which a belt transfers the driving power to the mill vial, has been used and the maximum revolution speed of the mill vial is ~ 400 rpm (P-4 or P-5, Firtsich). In contrast, an ultra high-energy planetary ball mill with a gear wheel, which transfers the driving power, can operate with approximately 670 rpm of the revolution speed (BX284EH, Kurimoto).

In this research, the improvement of the hydrogen storage properties of MgH_2 by the mechanical grinding (MG) process with an ultra high-energy planetary ball mill was investigated. The advantages, of using an ultra high-energy planetary ball mill, are having a wider range and a larger maximum value of the revolution speed. The former advantage makes it possible to investigate the influence of the milling condition on the hydrogen storage properties of MgH_2 by changing the revolution speed. The latter advantage has potential to lead to the reduction of MG time. Also, 50wt% of Al_2O_3 was added to the MgH_2 during MG process. This oxide may play a role in reducing the grain size, and has the potential to prevent grain growth at operating temperature.

EXPERIMENTAL PROCEDURE

The starting powder was MgH_2 (purity: 95%, particle size: -100 μm, Alfa Aesar) and α-Al_2O_3 (purity: 99.9%, average particle size: 2-3 μm, Kojundo Chemical Laboratory CO., LTD.,

Japan). The mixture of 50wt% Al_2O_3 and MgH_2 was ball-milled as MG process by ultra high-energy planetary ball mill (High-G BX284EH, KURIMOTO, Japan) with a gear wheel, which transfers a driving power to a mill vial. This mill exhibited the maximum revolution per minute of 670 rpm and fixed rotation / revolution speed ratio of 1.092 / 1. ZrO_2 balls and mill vials (powder / ball weight ratio = 1 / 40, ball weight: 185 g) were selected. MG time was 10min for all samples. All handling processes of the powders were performed in a glove box filled with Ar atmosphere (oxygen concentration $\leqq$ 3 ppm, water concentration $\leqq$ 24 ppm) to avoid the oxidation of the samples.

We used a volumetric Siverts apparatus (SUZUKI SHOKAN CO., LTD., Japan) in order to measure hydrogen storage properties. For the hydrogen desorption profile measurement, the samples were placed into a cell under 0.001 MPa of hydrogen as a starting pressure, and heated from room temperature to 350 °C with a heating rate of 5 °C/min. Initial hydrogen pressures of 8.2 and 0.16 MPa were selected to measure hydrogen sorption and desorption kinetics, respectively. H_2 pressure data was collected during all hydrogen storage property measurements, and compared with the blank measurement data to calculate the H_2 capacity of MgH_2.

The phase identification of powder samples was carried out using X-ray diffraction (XRD) analysis (RINT2500, Rigaku, Japan) with Cu Kα radiation operated at 200 mA and 20 kV. The particle properties were examined by Scanning Electron Microscopy (SEM:JSM-7401F) and Energy-Dispersive X-ray Spectroscopy (EDS:JED-2300F) and Transmitting Electron Microscopy (TEM: JEM-2010) observation. Focused Ion Beam method sliced the large particle of sample with approximately 20 μm for TEM analysis.

RESULTS AND DISCUSSION

Generally, planetary ball mill consisted of a turntable with an angular speed of revolution ω_1 and mill vials rotating around a turntable center with an angular speed ω_2. According to Zhao et al., the maximum acceleration exerted on a mill charge is estimated as,

$$a_{\max} = \left(\omega_1^2 / 2 \right) \left[L + D(1+R)^2 \right] \tag{1}$$

where L is the diameter of the orbit revolution, D is the inner diameter of a mill vial and R is the ratio of ω_2 / ω_1 [3]. This acceleration corresponds with the intensity of milling. When the powder receives the energy from balls, large acceleration and revolution speed of the mill vial indicate the strong intensity of this energy [3-6]. In this research, all milling conditions except the revolution speed of the mil vial were fixed, because changing other milling conditions (such as ω_2 / ω_1 and a ball filling fraction in the mill vial) also affects the motion of balls in the mill vial and milling mechanism [3-6].

Figure 1 shows X-ray diffraction patterns of MG-MgH_2-50wt% Al_2O_3. The samples were milled by ultra high-energy planetary ball mill with three different revolution speeds of the mill vial. The revolution speeds were 303 and 673 rpm, which correspond to 30 and 150G, respectively. MgH_2 and 50wt% Al_2O_3 were mixed by an agate mortar as shown as 0 G in Fig. 1. The comparison between the peaks corresponded to MgH_2 and Al_2O_3 suggested that the milling with 30 G for 10 min slightly changed the intensity ratio of these peaks. In contrast, the significant change on the MgH_2 peaks was observed after the MG process for 10 min with 150 G. The large intensity of milling might cause the introduction of a defect in MgH_2 crystal lattice and a reduction in grain size. Furthermore, the milling with 150 G for 10 min was able to lead the phase transition from β to γ-MgH_2, which starts the appearance under the pressure of 2.5 GPa, and co-exists with β-MgH_2 up to a pressure of 8 GPa [7,8]. Therefore, we could observe extremely broad peaks and high pressure phase of MgH_2.

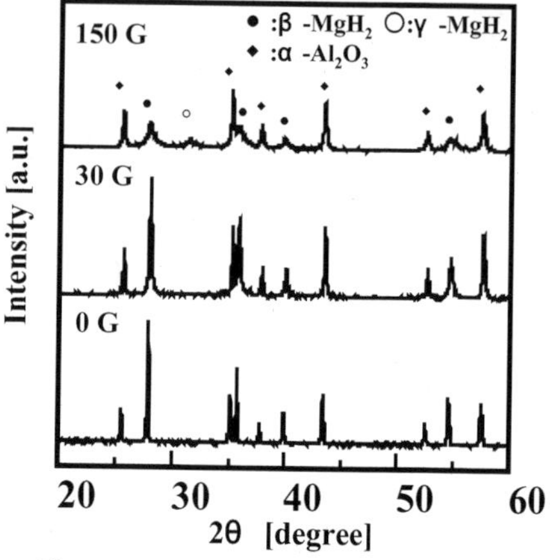

Figure1. X-ray diffraction patterns of MG-MgH_2-50wt% Al_2O_3.

Figure 2 shows the hydrogen desorption profiles of MgH_2 and MG-MgH_2-50wt% Al_2O_3 milled with 30 and 150 G. In the conditions of this research (such as a heating rate of 5 °C/min), the dehydriding reaction in untreated MgH_2 was not observed at the temperature lower than 350 °C due to its slow kinetics. Owing to the improvement of kinetics by the milling with 30 G for 10 min, the onset of H_2 capacity change was observed at 290 °C. However, the increase of the milling intensity from 30 to 150 G significantly affected to the onset of dehydriding reaction temperature, and decreased that down to 220 °C. Recently, the research of Mg-based system was reported that the onset of a dehydriding reaction temperature of MgH_2-17wt% Nb_2O_5 was decreased from 410 to 210 °C by MG

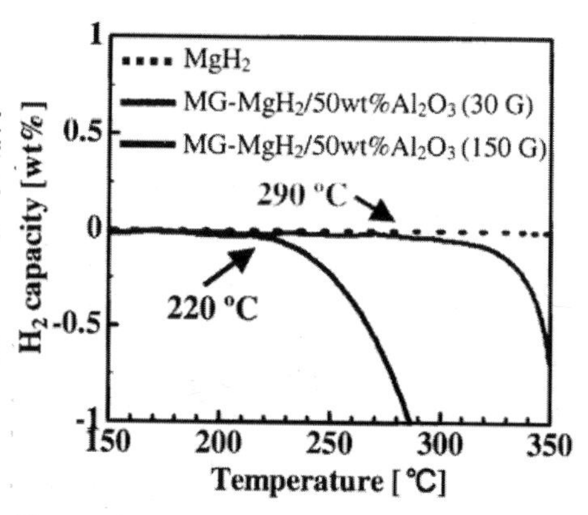

Figure 2. Hydrogen desorption profiles of MgH_2 and MG-MgH_2-50wt% Al_2O_3 milled with 30 and 150 G.

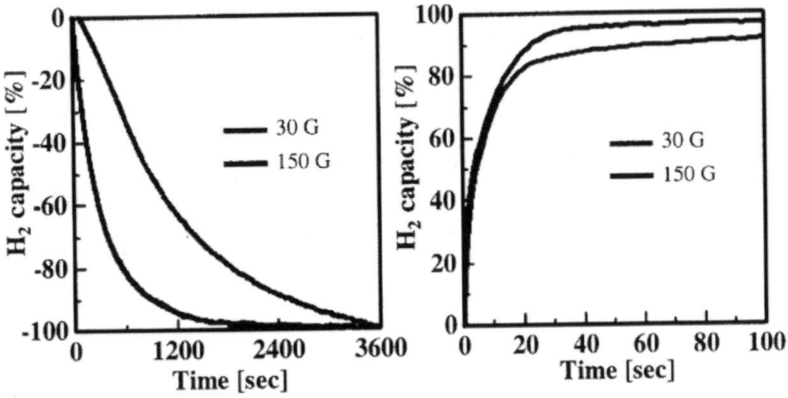

Figure 3. Hydrogen desorption kinetics of MG-MgH_2-50wt%Al_2O_3 at 350°C.

Figure 4. Hydrogen absorption kinetics of MG-MgH_2-50wt%Al_2O_3 at 350°C.

process for 200 h [2]. In contrast, the increase of the milling intensity caused the decrease in the dehydriding reaction temperature and the increase in the kinetics through 10 min milling.

Figure 3 and 4 show the kinetics of hydrogen desorption and absorption. In Fig. 3, the H_2 capacity change has the clear influence of the milling intensity. It is apparent that MG-MG-MgH_2-50wt% Al_2O_3 milled with 150 G for 10 min desorbed hydrogen significantly faster than the sample milled with 30 G. For example, the former sample required only 1/3 of time to desorb 80 % of hydrogen compared with the latter. After the desorption kinetics were measured, the sample became metal Mg, and then that reacted with 8.2 MPa of hydrogen as the initial pressure. The influence of the milling intensity on hydriding kinetics was relatively small as seen in Fig. 4. The difference in the time for completing 80% of hydriding reaction was only a few seconds between the two samples. The milling intensity showed the relatively un-clear effect in hydriding reaction, because the initial pressure was too high for 0.64 MPa, which is the equilibrium pressure of hydriding/dehydriding for Mg-MgH_2 at 350°C.

Figure 5 shows SEM and EDS mapping images of MG-MgH_2-50wt% Al_2O_3 milled with 150 G for 10 min. Before MG process, the particle size of un-treated MgH_2 was roughly 100 μm. After the milling process with 150 G, the sample consisted of very small and large particles with wide size distribution. Large particles exhibited flake-like shape due to the compaction between the balls and inner wall of the milling vial. Additionally, aggregated grains were observed from EDS mapping image, which showed that Mg and Al were homogeneously distributed in the same particle. This result suggested that the milling with 150 G for 10 min produced the composite of MgH_2 and Al_2O_3 with homogeneous distribution of those powders in micrometric level.

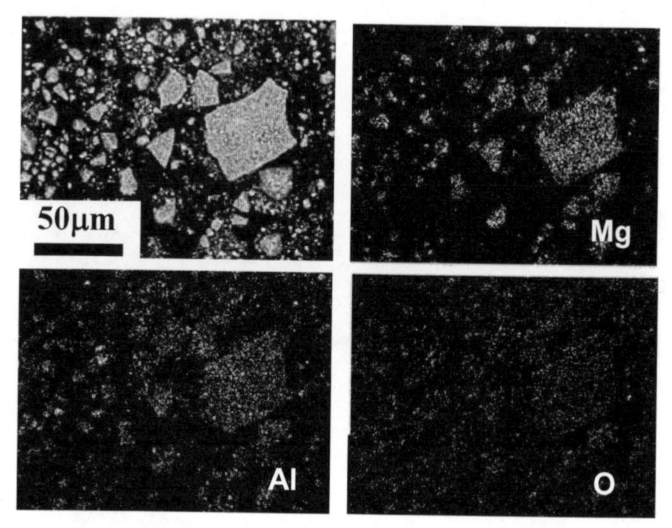

Figure 5. SEM and DES mapping images of MG-MgH_2 -50wt% Al_2O_3 milled for 10 min with 150 G.

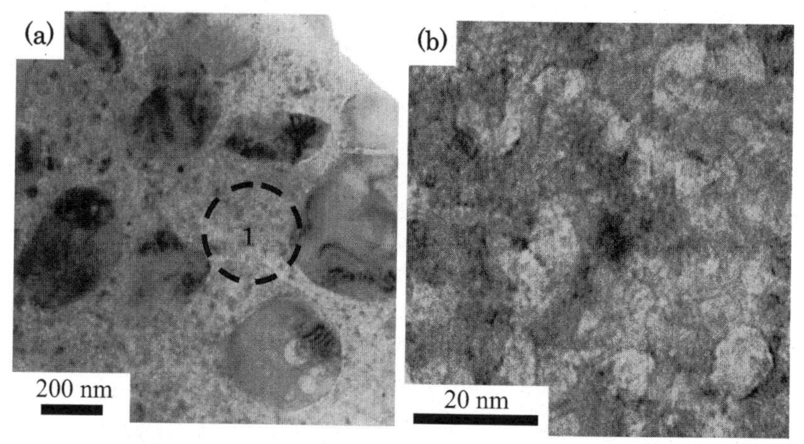

Figure 6. TEM images of MG-MgH_2 -50wt% Al_2O_3 milled for 10 min with 150 G. (a): Bright filed image. (b) Magnificated image taken from the region 1.

Figure 6 and 7 show TEM and EDS mapping images of MG-MgH_2-50wt% Al_2O_3 milled with 150 G for 10 min. The large aggregated grain with approximately 20 μm (likely as seen in Fig. 5) was selected to take this TEM image after slicing by FIB method. TEM image shows large grains surrounded by a matrix. The size of large grains was 150 to 400 nm. On the other

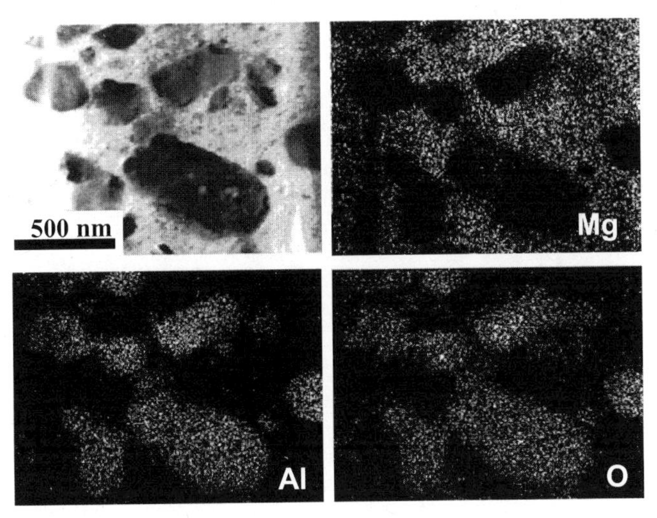

Figure 7. TEM and EDS mapping images of MG-MgH_2 /50wt%Al_2O_3 milled for 10 min with 150 G.

hand, remarkably small grains, which had the particle size less than 20 nm, existed in the matrix. From EDS mapping image, the distribution of MgH_2 and Al_2O_3 became obvious. Both powders co-existed corresponding to XRD result as seen in Fig. 1, in the aggregated grain as seen in Fig. 5. Also the large grain was Al_2O_3, and the matrix is mainly MgH_2. Repeated mechanical deformation during MG process decreases particle size of MgH_2. This phenomenon was enhanced by the combination of Al_2O_3 addition to MgH_2 and especially using strong milling intensity of 150 G. Therefore, MgH_2 consisted of small particle size of down to tens of nano-meters although the Al_2O_3 remained relatively large particle size after MG process for 10 min.

The drastic improvement of the hydriding/dehydriding kinetics was observed in Fig. 2, 3 and 4. This enhancement in the kinetics after ball milling attributed to a nanocrystalline microstructure with an increased volume of grain boundary, which act as a preferred pathway for hydrogen, and a reduced grain size, which decreases the mean diffusion path of hydrogen. Conventional planetally ball mills are possible to achieve supply 30 to 50 G and require long time to improve that kinetics [2,4]. In contrast, the providing 150 G by the ultra high-energy planetary ball mill leaded the drastic enhancement of sorption/desorption properties of hydrogen storage material.

CONCLUSION

We investigated the improvement of hydrogen storage properties of MgH_2 by the mechanical grinding (MG) process with an ultra high-energy planetary ball mill. The advantages of using ultra high-energy planetary ball mill are having a wider range and a larger maximum value of the revolution speed. The acceleration exerted on a mill charge corresponds with the intensity of milling. By changing the revolution speed, MgH_2-50wt% Al_2O_3 was milled as MG process with 0, 30 and 150 G for 10 min. XRD result showed the clear influence of the acceleration on the MgH_2 phase. The onset of H_2 capacity change was observed at 290 °C after MG process with 30 G. However, the increase of the milling intensity to 150 G decreased that down to 220 °C. The increase of the milling intensity significantly enhanced the hydriding/dehydriding kinetics of MgH_2. As the result of microstructure observation, the sample milled with 150 G for 10 min exhibits small particle size of down to tens of nano-meters. The drastic enhancement in the hydriding/dehydriding kinetics after ball milling attributed to a nanocrystalline microstructure. The influence of increasing the milling intensity was significant. The providing 150 G by the ultra high-energy planetary ball mill leaded the drastic enhancement of sorption/desorption properties of hydrogen storage material for a short time. Those results could be the answer for the request to obtain a system with efficiency and productivity in order to prepare hydrogen storage materials and improve their hydrogen storage properties.

ACKNOWLEDGEMENT

The authors acknowledge the partial support of this work by grant based on High-tech Research Center Program for private Universities from the Japan Ministry of Education, Culture, Sport, Science and Technology (MO).

REFERENCES

[1] G. Barkhordarian, T. Klassen, and R. Bormanna, Catalytic Mechanism of Transition-Metal Compounds on Mg Hydrogen Sorption Reaction, J. Phys. Chem. B, **110**, 11020 – 11024 (2006).

[2] K. F. Aguey-Zinsou, J.R. Ares Fernandeza, T. Klassena and R. Bormanna, Effect of Nb_2O_5 on MgH_2 properties during mechanical milling, Int. J. Hydrogen Energy, **32**, 2400-2407 (2007).

[3] Q. Q. Zhao, S. Yamada, and G. Jimbo, The mechanism and limit of fineness of a planetary mill grinding, J. Soc. Powder Technol. Jpn., **25**, 297-302 (1988).

[4] T. Takeuchi, K. Kaneko and U. Mizutani, Formation of Metastable Phases Using a High Energy Planetary Ball Mill, Kurimoto Technical Report, **28**, 28-33 (1993).

[5] H. Mio, J. Kano, F. Saito, and K. Kaneko, Effect of Rotational Direction and Rotation-to-Revolution Speed Ratio in Planetary Ball Milling, Mater. Sci. Eng. A, **332**, 75 – 80 (2002).

[6] B. Burgio, A Iasonna, M. Magini, S. Martelli, and F. Padella, Mechanical Alloying of the Fe-Zr System. Correlation between Input Energy and End Products, Il Nuovo Cimento, **12**, 459-476 (1991).
[7] F. C. Gennari, F. J. Castro, G. Urretavizcaya, Hydrogen desorption behavior from magnesium hydrides synthesized by reactive mechanical alloying, J. Alloys. Comp. **321**, 46-53 (2001).
[8] M. Bortz, B. Bertheville, G. Böttger, K. Yvon, Structure of the high pressure phase γ-MgH2 by neutron powder diffraction, J. Alloys. Comp, **287**, L4-L6 (1999).

AB INITIO STUDY OF THE INFLUENCE OF PRESSURE ON THE HYDROGEN DIFFUSION BEHAVIOR IN ZIRCONIUM HYDROGEN SOLID SOLUTION

Y. Endo, M. Ito, H. Muta, K. Kurosaki, M. Uno, S. Yamanaka
Division of Sustainable Energy and Environmental Engineering, Graduate School of Engineering, Osaka University
2-1 Yamadaoka, Suita, Osaka 565-0871, Japan.

ABSTRACT

We studied stress dependence of hydrogen diffusion behavior in zirconium by using density functional theory. Hydrogen solution energy was examined as a function of stress and we found that tensile stress leads to increase in hydrogen solution energy because it induces lattice expansion. Similarly, stress promotes increasing of partial molar volume of hydrogen and anisotropic stress along z direction has the largest impact on it. From the investigation of potential energy curves of hydrogen diffusion, we investigated activation energy and diffusion constant from T site to O site and O site to O site, respectively. The activation energy for hydrogen diffusion is decreased by stress, which induces lattice expansion perpendicular to diffusion path. Diffusion constant is increased by stress, which increases diffusion path length.

INTRODUCTION

Zirconium alloys are used as the fuel cladding materials in light water reactors (LWR) due to its low thermal neutron adsorption. During operation, the claddings get corroded with coolant water as follows: $\mathrm{Zr} + 2\mathrm{H_2O} \rightarrow \mathrm{ZrO_2} + 2\mathrm{H_2}$. As the result, hydrogen is absorbed onto the claddings. When the hydrogen concentration in zirconium exceeds the terminal solid solubility, zirconium hydride is precipitated. The mechanical properties of metal are changed greatly due to hydrogen, and therefore zirconium hydride is concern about its mechanical and chemical properties.

Recently, the burn up of LWR fuels has been increased in order to improve economy. However, in the power ramp test, the high burn up claddings failed during the test [1]. The failure mechanism are supposed that stress concentrates at initial crack tip of claddings, hydrogen diffuses to the tip, hydride precipitates and cracks, and then the crack tip propagates [1, 2]. The process is generally called delayed hydride crack (DHC).

As above, studying hydrogen diffusion mechanism is one of the most important countermeasures for comprehending DHC. The diffusion process depended on stress was studied by experimental and theoretical methods [3, 4]. Stress dependence of hydrogen diffusion coefficient in niobium was studied by Koike [5] and they found stress increases diffusion coefficient.

Hydrogen diffusion flux J is described by Fick's law:

$$J = -\frac{C_{\mathrm{H}} D}{RT} \cdot \left(R \ln\left(\frac{C_{\mathrm{H}}}{S_{\mathrm{H}}}\right) \cdot \frac{\partial T}{\partial x} + RT \cdot \frac{1}{C_{\mathrm{H}}} \cdot \frac{\partial C_{\mathrm{H}}}{\partial x} + \overline{V}_H \cdot \frac{\partial P}{\partial x} \right)$$

where C_H is hydrogen concentration, D is diffusion coefficient, S_H is hydrogen solubility, x is distance, and $\overline{V}_H$ is partial molar volume of hydrogen. In this formula D, S_H, and $\overline{V}_H$ may change with the stress. However, stress dependence of these parameters has not been examined both experimentally and theoretically yet.

First principles calculations based on density functional theory (DFT) [6, 7] have been applied for studying various properties such as solution energy and phase stability. The parameters of zirconium alloys have been studied extensively with the equilibrium lattice parameter. However, it has been previously shown that stress affects significantly on adsorption energy of transition metal surfaces such as palladium [8, 9, 10].

In this study, to evaluate stress dependence of diffusion behavior, stress dependence of solution energy and solubility, partial molar volume, and diffusion coefficient of hydrogen in zirconium was investigated using the first principle DFT calculations.

THEORETICAL METHOD

DFT calculations were performed by Cambridge Serial Total Energy Package (CASETP) [11] using a plane wave basis set. We used ultra-soft pseudo-potential [12] and the generalized gradient approximation of the Perdew-Wang 91 (GGA-PW91) [13] for the exchange-correlation energy.

The plane wave basis set cutoff energy was chosen to be 390 eV. The Monkhorst-Pack scheme [14] was used for the k-point sampling with a $9 \times 9 \times 5$ k-mesh for the primitive cell and $4 \times 4 \times 5$ k-mesh for the $2 \times 2 \times 1$ super cell. The electronic occupancies were determined with an energy smearing of 0.1 eV.

To confirm that our calculations had converged, we checked the convergence of energy with respect to cutoff energy, k-point sampling. Changing the number of k-points to $11 \times 11 \times 7$ k-mesh for primitive cell, the total energy changed by less than 11 meV. Changing the number of cutoff energy to 405 eV, the total energy changed by less than 0.4 meV. From these results, we confirmed that those parameters used in this study gave a convergence within 11 meV.

Solution energy was calculated by allowing all Zr and H atoms to completely relax (volume + atomic relaxations). The convergence criteria for energy charge, maximum force, maximum stress, maximum displacement, and maximum number of geometry optimization cycles were 5.0×10^{-06} eV/atom, 0.01 eV/Å, 0.02 GPa, 5.0×10^{-04} Å, 100, respectively.

Stress was introduced isotropically and anisotropically by changing the calculated lattice parameter. This approach has been previously shown to evaluate the effects of strain on transition metal reactivity [8, 9, 10]. In this study, stress magnitude represents equivalent hydrostatic pressure. A positive value of stress indicates that the structure is compressed.

We calculated the equilibrium bond length of the isolated H_2 molecule in $5 \times 5 \times 10$ Å^3 box. The results are summarized in Table I and compared with previous calculations and experiments. From these results, we obtained 0.753 Å for the hydrogen bond length. Our results agree well with the previous theoretical values.

Table I. Bond Length, Bonding Energy of Hydrogen

	Bond length Å
This study	0.753
Experiment [15]	0.742
Domain [16]	0.760

RESULT AND DISCUSSION

Hydrogen solution energy in zirconium

The calculated lattice parameters of zirconium are shown in Table II as a function of stress, which compared with the experimental [17] values and Domain's calculated values [18]. Our calculated lattice parameters were a = 3.222 Å, c = 5.189 Å and c/a = 1.611, which is larger than that of the experimental value of c/a = 1.604. This slight difference is owing to calculation methods such as wave function and the exchange correlation energy. As the stress changed, the lattice parameters changed to a = 3.217 Å, c = 5.184 Å at stress = 0.4 GPa and to a = 3.226 Å, c = 5.195 Å at stress = -0.4 GPa, respectively. When stress was applied along x-y direction (parallel to basal plane), the lattice parameter a changed larger than those of isotropic stress. Stress along z direction (perpendicular to basal plane) changed the lattice parameter c larger than those of isotropic stress.

Table II. Lattice Parameters of Zirconium

		Lattice parameter Å		
	Stress GPa	a	c	c/a
This study	0	3.222	5.189	1.611
	0.4	3.217	5.184	1.612
	-0.4	3.226	5.195	1.610
	0.4 (x-y)	3.210	5.205	1.622
	-0.4 (x-y)	3.234	5.174	1.600
	0.4 (z)	3.231	5.141	1.591
	-0.4 (z)	3.212	5.240	1.632
Experiment [17]		3.232	5.148	1.593
Domain [16]		3.230	5.180	1.604

It is well known that the hydrogen solubility in pure zirconium follows Sieverts' law: $S_H = K_H P_{H_2}^{1/2}$ where S_H is the hydrogen solubility expressed as an atomic ratio (H/Zr), P_{H2} is the equilibrium hydrogen pressure and K_H is the Sieverts constant. The Sieverts constant as function of temperature is described by an Arrenius type equation: $\ln K_H = \frac{\Delta S}{R} + \frac{\Delta H}{RT}$ where ΔS is the entropy of solution and ΔH is the enthalpy of solution. ΔS is related configurational entropy, vibrational entropy and electronic entropy and we used the experimental value ΔS = -11.9 [18] because we could not calculate it using CASTEP. The enthalpy of solution ΔH (the solution energy E_{sol}) is defined as:

$$E_{sol} = E_{Zr-H} - E_{Zr} - \frac{1}{2}E_{H2}$$

where $E_{Zr\text{-}H}$ is the total energy of primitive cell of Zr-H, E_{Zr} is the total energy of a similar primitive cell not including hydrogen atom, and E_{H2} is the total energy of a H_2 molecule in a cube with dimensions of 5 × 5 × 5 Å^3. We considered Hydrogen occupied tetragonal site [19].

The calculated solution energies as a function of stress are showed in Table III compared with experimental [14] and Domain's value [16]. In the case of no stress, the solution energy was -

0.57 eV, which is smaller than experimental value (-0.66 eV) and Domain's value (-0.604 eV). In the case of 0.4 GPa (compressive stress), the solution energy was smaller than no stress, whereas in the case of -0.4 GPa (tensile stress), the solution energy was larger than no stress. We could not find anisotropic stress dependence of solution energy contrary to other parameters as shown below. From these results, we calculate hydrogen solubility change using Sieverts' law and suggest that the hydrogen solubility (H/Zr) should change -43.0 % GPa^{-1} as a function of isotropic stress at 573 K.

The results indicate that the solution energy depends on stress and tensile stress increases the solution energy. This dependence agrees well with previous theoretical study [8, 9, 10]. This results show that hydrogen is stable in the tensile zirconium, which is explained by that hydrogen atom needs a large lattice expansion around interstitial site [13] and tensile stress promotes the expansion.

Table III. Hydrogen Solution Energies Depended on Stress

	Stress GPa	E_{sol} eV
This study	0	-0.570
	0.4	-0.561
	-0.4	-0.578
Experimental [14]	0	-0.660
Domain [16]	0	-0.604

Partial molar volumes of hydrogen

In Zr-H system, the volume expansion has been investigated experimentally and theoretically in several studies [16, 20]. The results of stress dependence of partial molar volumes (relaxation volumes) are displayed in Table IV. In this study volume expansion was 3.389 $Å^3$/atom, which is larger than experimental value of 2.8 $Å^3$/atom. In the case of 0.4 GPa, we found 0.006 $Å^3$/atom decrease compared to the corresponding equilibrium lattice parameter, whereas in the case of -0.4 GPa, we found 0.008 $Å^3$/atom increase compared to isotropic stress. From these results, we suggest partial molar volume of hydrogen should change -0.51 % GPa^{-1} as a function of isotropic stress.

Besides this, we found anisotropic stress dependence of relaxation volume. The effect at stress along *x-y* direction was lower than isotropic stress, whereas the effect at stress along *z* direction was larger than isotropic stress. This indicates that stress dependence is anisotropic and stress impact along *z* direction is strong.

Table IV. Partial Molar Volumes of Hydrogen Depended on Stress

	Stress GPa	V $Å^3$/atom	ΔV $Å^3$/atom
This study	0	3.389	0.000
	0.4	3.384	-0.006
	-0.4	3.398	0.008
	0.4 (*x-y*)	3.385	-0.005
	-0.4(*x-y*)	3.391	0.001
	0.4 (*z*)	3.383	-0.007
	-0.4 (*z*)	3.402	0.013
Experimental [20]	0	2.800	-
Domain [16]	0	3.900	-

Diffusion coefficient

In this section, to calculate the activation energy for hydrogen diffusion in zirconium, we adopted 2 × 2 × 1 supper cell containing 8 zirconium atoms. The cell was fixed (volume + atomic fix) because relaxation calculation cost is fairly high. Lattice parameters used in this section were already showed in Table II.

In the hexagonal close-packed (hcp) α-Zr, there are two high-symmetry interstitial sites available for hydrogen atom occupation: the octahedral site (O site) and the tetrahedral site (T site). For examination of hydrogen diffusion, we chose 4 T sites and 3 O sites showed in Figure 1. We investigated hydrogen diffusion between neighbor stable sites. In these site, two diffusion path is occurred [16]: the first path is $T_1 \rightarrow O_1$ asymmetric path (T-O diffusion path), and the second path is $O_1 \rightarrow O_2$ symmetric path (O-O diffusion path). The other paths have significantly high activation energy for diffusion, so we ignored those paths.

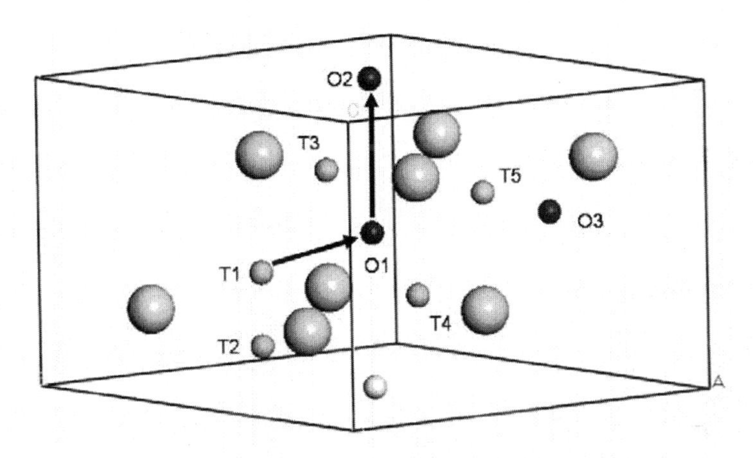

Figure 1. Hydrogen Diffusion Pathes in Zirconium-Hydrogen

The potential energy curves (PECs) for hydrogen in zirconium at no stress are shown in Figure 2. Activation energy for diffusion can be calculated as follow: $E_a = \Delta E + E_0$, where E_a is activation energy corrected with the zero- point energy, ΔE is energy difference between the saddle point and ground state on the potential energy surface, and E_0 is zero point energy of hydrogen. The zero-point energy of hydrogen is calculated by the following:

$$E_0 = \frac{1}{2} h \nu_0$$

where ν_0 is vibration frequency of hydrogen. The vibration frequency is calculated as curvature of hydrogen to displace 0.1 Å from the ground state position to transition position along the diffusion path by following:

$$\nu_0 = \frac{1}{2\pi}\sqrt{\frac{2k}{m}}$$

where m is hydrogen mass, and k is curvature of diffusion path [21].

We found that the activation energy for diffusion from T site to O site and O site to O site was 0.394 eV, 0.487 eV, respectively. From table V, it was indicated that as the stress increased the activation energy increased proportionately. In the case of T-O diffusion path, stress along the x-y direction had less effect on activation energy compared with isotropic stress, and stress along the z direction had more effect on it. On the other hand, in the case of O-O diffusion path, stress along x-y direction had more effect compared with isotropic stress, and stress along z direction had less effect on it. These trends result from the fact that because lattice expansion is required for hydrogen to migrate to neighbor site [22] hydrogen diffuses more easily in expansion lattice than equilibrium lattice. Due to this it is suggests that stress, which expands lattice perpendicular to diffusion path, leads to enhance hydrogen diffusion in zirconium.

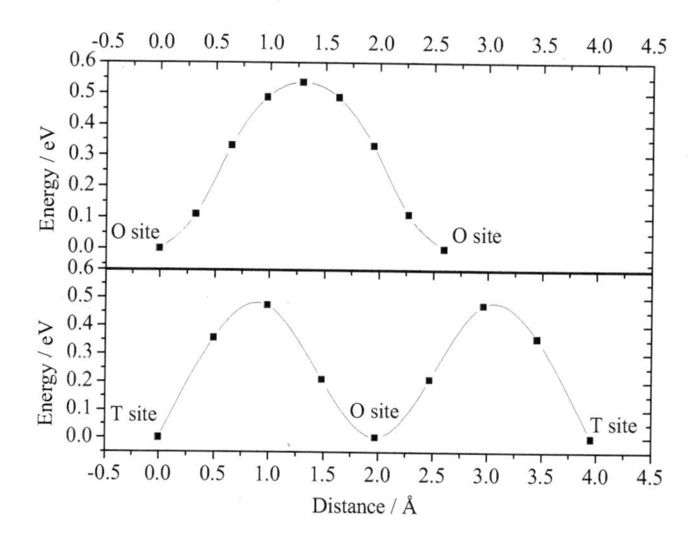

Figure.2 Potential Energy Curve of Hydrogen Diffusion in Zirconium

Table V. Activation Energy Depended on Stress

	stress GPa	Activation energy eV						
		0	0.4	-0.4	0.4 (*x-y*)	-0.4(*x-y*)	0.4 (*z*)	-0.4 (*z*)
This work	T-O	0.394	0.395	0.393	0.393	0.394	0.398	0.389
	O-O	0.487	0.497	0.477	0.500	0.475	0.493	0.481
Domain[16]	T-O	0.350	-	-	-	-	-	-
	O-O	0.410	-	-	-	-	-	-
Experiment [23]	T-O	0.385	-	-	-	-	-	-
	O-O	0.415	-	-	-	-	-	-

The temperature dependence of diffusion coefficient D follows the Arrenius law:

$$D = D_0 \exp\left(-\frac{E_a}{RT}\right)$$

where D_0 is the diffusion constant. The diffusion constant D_0 is calculated as [21, 24]:

$$D_0 = \frac{1}{6} a^2 v_0$$

where a is the jump length. The calculated D_0 are shown in Table VI. The results were D_0= 2.49 m^2/s (T-O site) and 2.66 m^2/s (O-O site), respectively. When compressive stress was applied, diffusion constant was larger than that in the case of no stress. In the case of T-O diffusion path, stress along x-y direction had negative effect on diffusion constant, and stress along z direction had positive effect on it. On the other hand, in the case of O-O diffusion path, stress along x-y direction had positive effect on diffusion constant, and stress along z direction had negative effect it. This trend results from the fact that diffusion constant is sensitive to jump length a as shown above formula. Due to this, it is indicate that stress, which strains parallel to diffusion path, leads to activate hydrogen diffusion in zirconium. From these results, we found the diffusion coefficient D should change -9.2 % GPa^{-1} as a function of isotropic stress at 573 K.

Table VI Diffusion Constants Depended on Stress

	stress GPa	diffusion constant 10^{-7} m^2/s						
		0	0.4	-0.4	0.4 (*x-y*)	-0.4(*x-y*)	0.4 (*z*)	-0.4 (*z*)
This work	T-O	2.49	2.49	2.48	2.46	2.52	2.56	2.42
	O-O	2.66	2.67	2.64	2.69	2.62	2.63	2.69
Domain [25]	T-O	6.7	-	-	-	-	-	-
	O-O	4.1	-	-	-	-	-	-
Experiment [23]	T-O	0.34	-	-	-	-	-	-
	O-O	1.73	-	-	-	-	-	-

CONCLUSION

In this study we investigated the effect of stress of hydrogen diffusion behavior and its origin by DFT calculations. We found that stress plays a major role in determining hydrogen diffusion flux in zirconium. There is a correlation between stress and solution energy, partial mole volumes, and diffusion coefficient. Tensile stress leads to increase hydrogen solution energy because it increases lattice parameter and hydrogen can be stable in expanded lattice. Partial molar volume of hydrogen is increased by tensile stress and anisotropic stress along z direction has the largest impact on it. We examined the potential energy curve of hydrogen diffusion in zirconium. Activation energy of diffusion is decreased by tensile stress because it induces lattice expansion perpendicular to diffusion path. Tensile stress increases diffusion constant due to its expanding diffusion path length.

REFERENCES

[1] S. Shimada, E. Etoh, H. Hayashi, Y. Tukuta, A metallographic and fractographic study of outside-in cracking caused by power ramp tests, *J. Nucl. Mater.*, 327, 97-113 (2004).

[2] H. Hayashi, K. Ogata, T. Baba, K. Kamimura, Research Program to Elucidate Outside-in Failure of High Burnup Fuel Cladding, *J. Nucl. Sci. Technol.*, 43, 1128-1135 (2006).

[3] Kim, Y.S., Ahn, S.B., Cheong, Y.M., Precipitation of crack tip hydrides in zirconium alloys, *J. Alloys Compd.*, 429, 221-226 (2007).

[4] A. Varias, A. Massih, Simulation of hydrogen embrittlement in zirconium alloys under stress and temperature gradients, *J. Nucl. Mater.*, 279, 273–285 (2000).

[5] S. Koike, A. Kojima, M. Kano, M. Otake, H. Kojima, T. Suzuki, On the Superdiffusion of Hydrogen in V_a-Metals, J. Phys. Soc. Jpn., 59, 584-595 (1990).

X S. Yamanaka, K. Higuchi, M. Miyake, Hydrogen solubility in zirconium alloys, *J. Alloys Compd.*, 231, 503-507 (1995).

[6] Hohenberg, P., Kohn, W., Inhomogeneous electron gas, *Phys. Rev.*, 136, B864-B871 (1964).

[7] Kohn, W., Sham, L.J., Self-consistent equations including exchange and correlation effects, *Phys. Rev.*, 140, A1133-A1138 (1965).

[8] Grabow L, Xu Y, Mavrikakis M., Lattice strain effects on CO oxidation on Pt(111), *Phys Chem Chem Phys.*, 8, 3369-74 (2006).

[9] M. Mavrikakis, B. Hammer, J. K. Nørskov, Effect of strain on the reactivity of metal surfaces, *Phys. Rev. Lett.*, 81, 2819 - 2822 (1998).

[10] Ozawa, N., Arboleda Jr., N.B., Nakanishi, H., Kasai, H., First principles study of hydrogen atom adsorption and diffusion on Pd3Ag(1 1 1) surface and in its subsurface, *Surf. Sci.*, 602, 859-863 (2008).

[11] Segall, M.D., Lindan, P.J.D., Probert, M.J., Pickard, C.J., Hasnip, P.J., Clark, S.J., Payne, M.C., First-principles simulation: Ideas, illustrations and the CASTEP code, *J. Phys.: Condens. Matter* 14 2717-2744 (2002).

[12] Vanderbilt, D., Soft self-consistent pseudopotentials in a generalized eigenvalue formalism, *Phys. Rev. B*, 41, 7892-7895 (1990).

[13] Perdew, J. P., Chevary, J. A., Vosko, S. H., Jackson, K. A., Pederson, M. R., Singh, D. J., Fiolhais, C., Atoms, molecules, solids, and surfaces: Applications of the generalized gradient approximation for exchange and correlation, *Phys. Rev. B*, 46, 6671-6687 (1992).

[14] Monkhorst, H. J., Pack, J. D., "Special points for Brillouin-zone integrations" - a reply, *Phys. Rev. B*, 16, 1748-1749 (1977).
[15] K. Christmann. Interaction of hydrogen with solid surfaces, *Surf. Sci.,* 9 1-163 (1988).
[16] Domain, C., Besson, R., Legris, A., Atomic-scale Ab-initio study of the Zr-H system: I. Bulk properties, *Acta Mater.*, 50, 3513-3526 (2002)
[17] H.E. Swanson, R.K. Fuyat., Standard X-Ray Diffraction Powder Patterns, *Nat. Bur. Stand. U.S.*, 539 II, 11(1953)
[18] S. Yamanaka, K. Higuchi, M. Miyake, Hydrogen solubility in zirconium alloys, *J. Alloys Compd.,* 231, 503-507 (1995).
[19] Khoda-Bakhsh, R., Ross, D.K., Determination of the hydrogen site occupation in the α phase of zirconium hydride and in the α and β phases of titanium hydride by inelastic neutron scattering , *J. Phys.* F: Metal Physics, 12,15-24 (1982).
[20] S.R. MacEwen, C.E. Coleman, C.E. Ells, J. Faber Jr., Dilation of h.c.p. zirconium by interstitial deuterium, *Acta Metall.*, 33, 753-757 (1985).
[21] Fukai, Y, The metal-hydrogen system, (1993).
[22] H. Dosch, F. Schmid, P. Wiethoff, J. Peisl, Lattice-distortion-mediated local jumps of hydrogen in niobium from diffuse neutron scattering, *Phys. Rev. B* 46, 55 - 68 (1992).
[23] C.S. Zhang, B. Li, P.R. Norton, The study of hydrogen segregation on Zr(0001) and Zr(1010) surfaces by static secondary ion mass spectroscopy, work function, Auger electron spectroscopy and nuclear reaction analysis, *J. Alloys Compd.* 231 354-363 (1995).
[24] H. Sugimoto, Diffusion Mechanism of Hydrogen in Solids, *Vac. Soc. Jpn.*, 49 17-22 (2006).
[25] Domain, C., Ab initio modelling of defect properties with substitutional and interstitials elements in steels and Zr alloys, *J. Nucl. Mater.*, 351, 1-19 (2006).

EBSP STUDY OF HYDRIDE PRECIPITATION BEHAVIOR IN Zr-Nb ALLOYS

Shunichiro Nishioka, Masato Ito, Hiroaki Muta, Masayoshi Uno, Shinsuke Yamanaka
Division of Sustainable Energy and Environmental Engineering, Graduate School of Engineering, Osaka University
2-1,Yamadaoka, Suita, Osaka 565-0871, Japan

ABSTRACT

Microscopic texture and precipitation behavior of zirconium hydrides in recrystallized Zircaloy-2, Zr-1.0 wt%Nb, Zr-2.5 wt%Nb and stress-relieved Zircaloy-4 have been examined by electron backscattering diffraction patterns (EBSP). The microscopic texture data of α-matrix grains obtained by EBSP were consistent with the macroscopic X-ray data. The α-matrix grains in Zr alloys have normal direction textures of (0001) basal planes. β-phase grains in Zr-1.0Nb and Zr-2.5Nb have rarely orientation. δ-hydride grains in all the specimens have normal direction textures of {001} basal planes. In this study, $\{10\bar{1}7\}_{\alpha}||\{111\}_{\delta}$ is more common hydride habit plane relationship than $(0001)_{\alpha}||\{111\}_{\delta}$. On the other hand, the third relationship $\{10\bar{1}1\}_{\alpha}||\{110\}_{\delta}$ was hardly observed in Zry-2, but often observed in Zr-1.0Nb, Zr-2.5Nb and Zry-4.

INTRODUCTION

The surface of fuel cladding tubes used in light-water reactors oxidizes through corrosion reaction with coolant water and hydrogen is generated during operation. The hydrogen is partly absorbed by the claddings [1,2]. The hydrogen absorption normally defines the life and performance of fuel cladding tubes in light-water reactors. Although pure Zr and Zr-rich alloys dissolve up to 450 ppm hydrogen in solid solution at around 773 K, the solubility decreases markedly as the temperature is lowered, with 65 ppm hydrogen at 573 K and 0.05 ppm at room temperature [3-5]. When the hydrogen quantity exceeds the hydrogen solubility limit, it precipitates as brittle zirconium hydrides. It is reported that the precipitation of zirconium hydride in zirconium alloys drastically reduces the ductility, fracture toughness and impact strength at room temperature as well as at reactor operating temperature [4,6].

Amount of hydrogen absorbed into claddings will increase with the burnup extension of light-water reactors. Therefore it is necessary to improve corrosion resistivity and mechanical properties of the cladding tube. Nb addition to Zr-based alloys is known to improve corrosion resistivity and mechanical properties and the Zr-Nb alloys are expected for the new cladding tube materials. Some Zr-Nb alloys, for example NDA, MDA and ZIRLO have been newly used in light-water reactors. Zr-Nb binary phase diagram is shown in Figure 1. Unlike Zircaloys which are mainly single α-phase (hcp structure) material, Zr-Nb alloys have the microstructure of the β-phase (bcc structure) distributed as stringers along the grain boundaries of the α-phase [7]. The precipitation of β-phase influences on not only the mechanical properties but also residual stress of Zr alloys. Therefore, hydride precipitation behavior in Zr-Nb alloys is expected to be different from that in Zircaloys. Hence, the formation of hydrides in Zr-Nb alloys was evaluated by EBSP measurement and the orientation and the habit planes were compared with those in Zircaloys in this study.

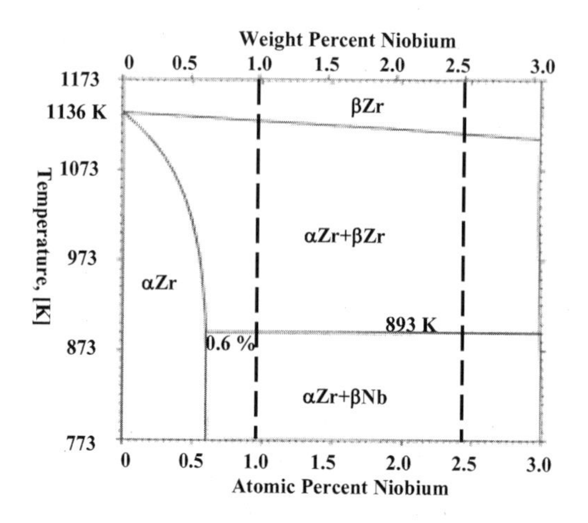

Figure 1. Zr-Nb binary phase diagram.

EXPERIMENTAL

The specimens of Zr-1.0Nb, Zr-2.5Nb, Zircaloy-2 and Zircaloy-4 were produced by Sumitomo Metal Industry Ltd. The preparation process of Zr-1.0Nb, Zr-2.5Nb and Zircaloy-2 was the same as that for the commercially supplied recrystallized Zircaloy-2. The preparation process of Zircaloy-4 was the same as that for the commercially supplied stress-relieved Zircaloy-4. The composition of these alloys was tabulated in Table I.

Table I. The composition of Zr-1.0Nb and Zr-2.5Nb (wt%)

Sample name	Fe	Sn	Cr	Nb	H	C	N	O	Zr
Zr-1.0Nb	0.002	0.038	0.002	1.00	0.0025	0.006	0.0047	0.100	Balance
Zr-2.5Nb	0.001	0.038	0.004	2.56	0.0032	0.008	0.0031	0.181	Balance

The hydrogenation of specimens was performed using a modified Sieverts' UHV apparatus at 723 K. Hydrogen contents of the Zr-1.0Nb and the Zr-2.5Nb specimens are about 250 wtppm. The phases of the specimens were analyzed by XRD.

The final surface treatment of polished samples for EBSP mapping was carried out by an ion-milling procedure. Crystallographic orientation measurements in specimens were made using a FE-SEM (JEOL-JSM6500F) equipped with an EBSP system (TSL-OIM ver.4.0). The estimated diameter of the electron beam at an acceleration voltage of15 kV and the scanning step adopted for EBSP mapping were 0.1 μm. We observed cross-section surface, which is perpendicular to rolling direction of specimens. The relationship of crystallographic orientation among the hydrides, β-phase and the surrounding α-Zr matrix grains was evaluated by normal pole figure plots of the Zr matrix and hydride phases at each analysis point.

RESULTS AND DISCUSSION

i) The microtexture

The Kikuchi patterns indexed by the EBSP system for all the specimens show that there are α-phase of the hcp structure, β-phase of the bcc structure and δ-ZrH_x of the fcc structure. Figure 2 (a)-(d) shows image quality (IQ) maps of specimens. IQ means definitions of Kikuchi patterns. Sharp Kikuchi patterns make IQ map clear while fuzzy Kikuchi patterns, for example those in grain boundaries, make IQ map unclear. The IQ maps of the recrystallized specimens: Zircaloy-2, Zr-1.0Nb and Zr-2.5Nb are relatively clear and every single grain can be distinguished. Meanwhile, the IQ map of the Zircaloy-4 was generally unclear. This was because Stress-relieved Zircaloy-4 had larger lattice strains or higher density of dislocations.

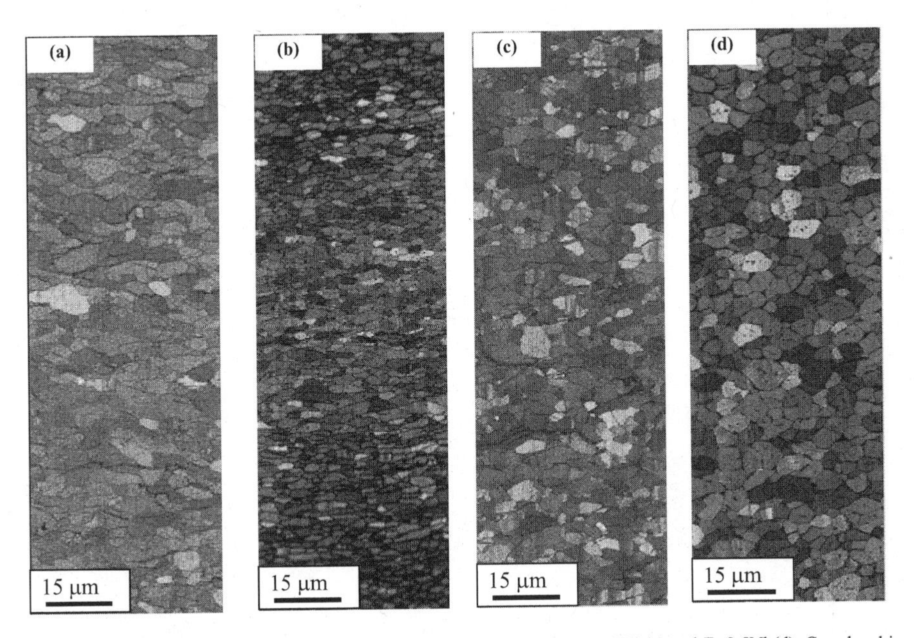

Figure 2. The image quality (IQ) maps of Zircaloy-2(a), Zircaloy-4(b), Zr-1.0Nb(c) and Zr-2.5Nb(d). Gray level in the maps represents relative intensity of the lattice strain.

Figure 3 shows examples of crystal direction maps of the specimens. In the maps, each α-Zr grain is categorized into three different gray levels, which represente the tilt angle ranges of the basal pole from the normal direction of specimens: (1) 0–30° , (2) 30–60° and (3) 60–90° . Lighter grains have more normal-direction orientations of the basal pole. Green grains in the maps indicated β-phase grains and yellow ones indicated δ-ZrH_x grains. These kinds of crystal direction mapping with a spatial resolution of sub-micron size for polycrystalline bulk specimens have been realized only by the recent EBSP technique.

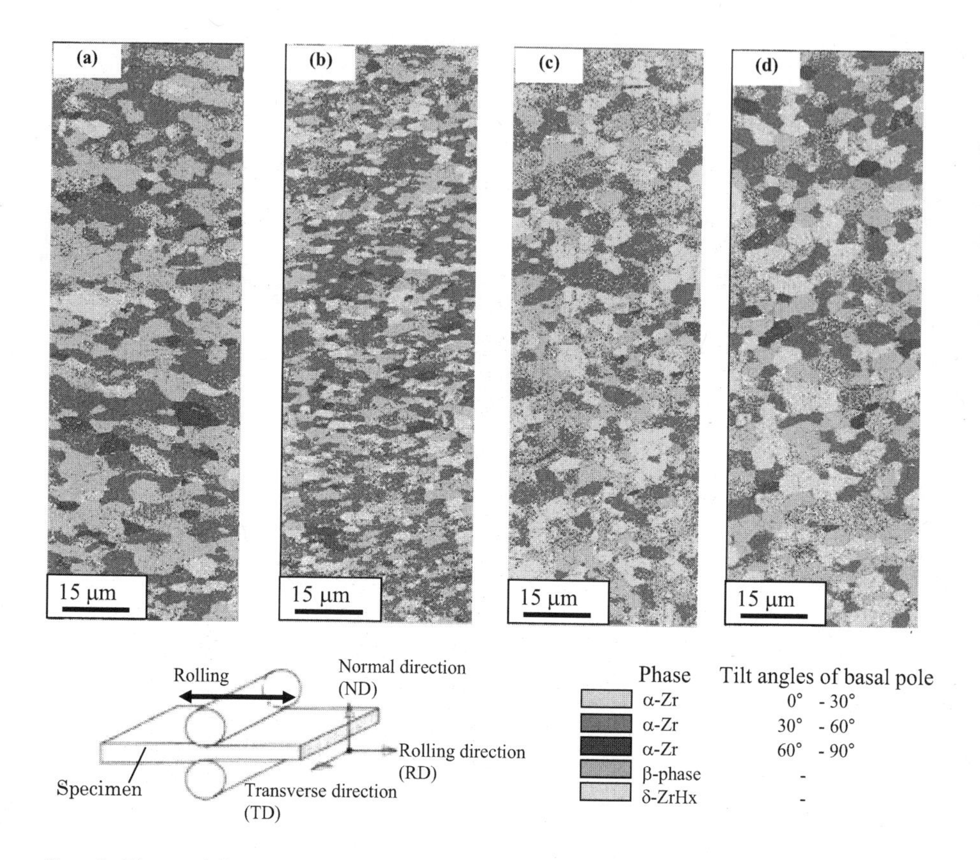

Figure 3. The crystal direction maps of Zircaloy-2(a), Zircaloy-4(b), Zr-1.0Nb(c) and Zr-2.5Nb(d). Three different gray levels of α-matrix grains indicate the tilt angles of the basal pole from the normal direction of the specimen. Green grains are β-phase grains. Yellow ones are δ-ZrH_x grains.

Comparing the grain size of α-matrix of the recrystallized Zircaloy-2 and stress-relieved Zircaloy-4, the latter specimen has smaller grains naturally. It is noted that most grains were oriented in the normal direction, as shown in Figure 3. This trend is closely related to the condition of cold rolling and annealing. In the case of tubing material, it is known that α-Zr matrix has a radial texture. The average grain diameters and the area fractions for α-matrix, β-phase and δ-hydride grains in the four specimens, which were calculated from the crystal direction maps, are summarized in Table II. The average grain diameter of Zircaloy-4 is smaller than others, as shown in Table II. This is because grains in stress-relieved Zircaloy-4 had not grown a lot during annealing.

Table II. The average grain diameters and the area fractions for α-matrix, β-phase and δ-hydride grains in the four specimens

	Average grain diameter [μm]				Area fraction [%]		
	α-matrix	β-phase	δ-hydride		α-matrix	β-phase	δ-hydride
Zr-1.0Nb	3.4	0.3	2.9		60	8	32
Zr-2.5Nb	3.9	0.6	3.0		63	8	29
Zircaloy-2	3.8	-	2.6		73	-	27
Zircaloy-4	2.2	-	2.0		81	-	19

The degree of macroscopic or average oriented texture of cladding tubes is generally given in terms of *f* value by X-ray diffraction measurements (XRD), which represents the fraction of all basal poles oriented in each direction [8]. Since the present EBSP system does not equip an evaluation software of *f* value, we adopt a normal direction texture parameter from the microscopic EBSP data, which was given as a percentage of the c-axis of α-Zr located within ±40° of the specimen normal direction. The evaluated values of this index were given in Table III.

Table III. Fraction of Zr grains whose basal poles are tilted within ±40° from the normal direction of the basal plane of the specimens

Zr-1.0Nb	Zr-2.5Nb	Zircaloy-2	Zircaloy-4	Zircaloy-2[a]
54 %	51 %	63 %	40 %	60 %

[a] Literature data of radial texture parameter in recrystallized tube of Zircaloy-2 [9]

The normal-direction texture parameter of recrystallized Zircaloy-2 (63 %) is certainly stronger than that of recrystallized Zr-1.0Nb (54 %) and Zr-2.5Nb (51 %). This indicates that the grains in Zr-Nb alloys did not have similar rotations during the recrystallization or grain growth process as those in Zircaloy-2 even though they had the same recrystallized annealing process.

Figure 4 shows the basal pole figure of α-matrix, β-phase and δ-hydride grains in the four specimens (|| normal direction), evaluated from the EBSP mapping data. The tilt angle of pole figures of α-matrix grains ranges from 0° to 90° . The tilt angle of pole figures of β-phase and δ-hydride grains ranges from 0° to 45° . The α-matrix grains in four specimens have normal direction textures of (0001) basal planes, as shown also in Figure 4. The random plots of pole figures of β-phase in Zr-1.0Nb and Zr-2.5Nb indicate that β-phase grains have no orientation. On the other hand, δ-hydride grains in four specimens have normal direction textures of {001} basal planes. This result is in agreement with previous data [10,11].

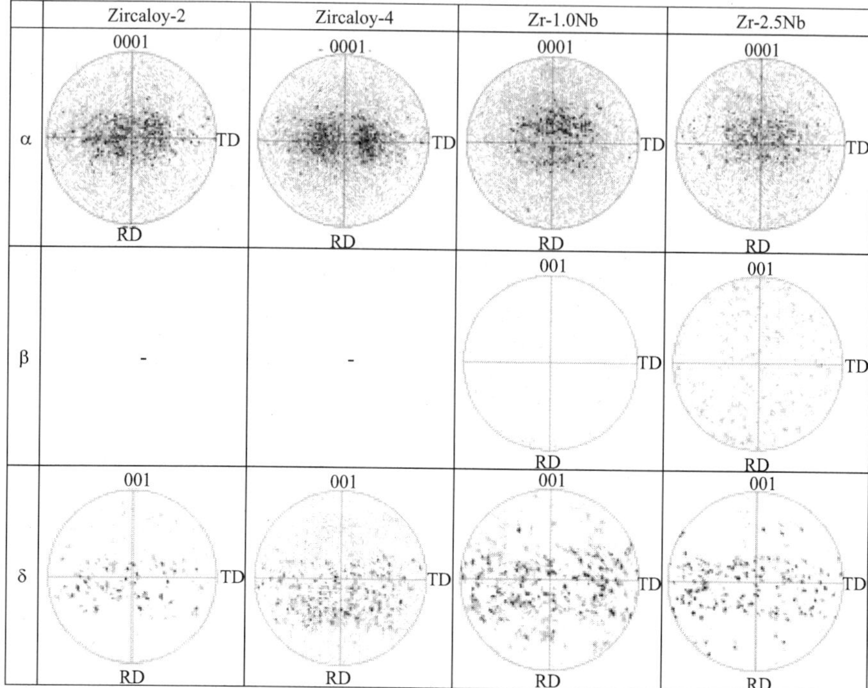

Figure 4. Basal pole figures of α-matrix, β-phase and δ-hydride grains in the four specimens (normal direction).

ii) The hydride habit plane

The crystallographic relationships between micron sized hydrides and the surrounding Zr grains for the four specimens were surveyed by means of the normal pole figure method [12]. The crystal planes examined were (0001), $\{10\bar{1}7\}$, $\{10\bar{1}1\}$ for the α-Zr matrix grains, and {111}, {110} for the δ-hydride grains. Assuming a tolerance angle of ±5° in a single crystal for the present EBSP data analyses, the coincidence between the crystallographic orientations of the δ-hydride grains and α-matrix grains was judged by a difference in angle of <±5° . The example of the pole figure coincidence between them is given in Figure 5. Figure 5 (a) is a part of phase map of hydrogenated Zr-2.5Nb evaluated from the EBSP mapping data. The red, green and yellow parts in the map mean α-matrix, β-phase and δ-hydride, respectively. Figure 5 (b) shows contiguous two grains in (α-matrix grain 1 and δ-hydride grain 2 in Figure 5 (a)) have the $\{10\bar{1}7\}_{\alpha}\|\{111\}_{\delta}$ relationship, clearly. In this study, randomly selected 25 groups of contiguous α-matrix grain and δ-hydride grains were analyzed. Table IV summarizes the result of the hydride habit plane in all the specimens.

Unlike the literatures [9,12-17], $(0001)_{\alpha}\|\{111\}_{\delta}$ relationship was hardly observed in all specimens. In this study, $\{10\bar{1}7\}_{\alpha}\|\{111\}_{\delta}$ is more common relationship than $(0001)_{\alpha}\|\{111\}_{\delta}$.

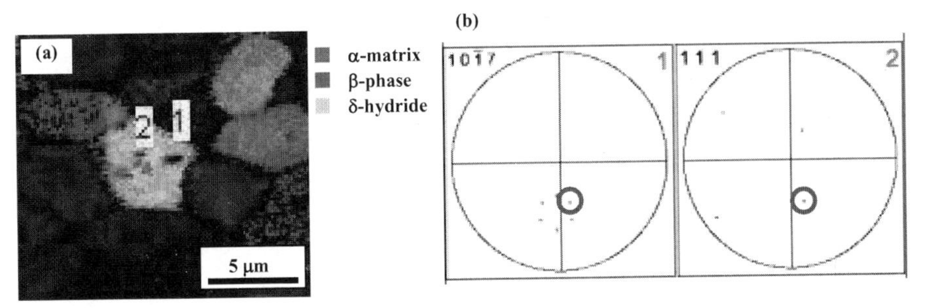

Figure 5. The example of the pole figure coincidence between α-matrix and δ-hydride.

The $\{10\bar{1}7\}$ plane is tilted by 15° from the (0001) basal plane, and has hexagonal symmetry. It is thought this result was being influenced by specimen shape, rolling condition, annealing condition and so on. On the other hand, the third relationship $\{10\bar{1}1\}_{\alpha}||\{110\}_{\delta}$ was hardly observed in Zrircaloy-2, but often observed in Zr-1.0Nb, Zr-2.5Nb and Zircaloy-4. $\{10\bar{1}1\}$ was reported to be observed only in stress-relieved Zircaloy [9]. In this study, $\{10\bar{1}1\}$ was often observed in stress-relieved Zircaloy-4 and hardly observed in recrystallized Zircaloy-2, exactly. However, $\{10\bar{1}1\}$ was observed also in recrystallized Zr-1.0Nb and Zr-2.5Nb. It is thought this result was being influenced by the residual stress.

Table IV. The observation frequency of the hydride habit plane in four specimens

	$(0001)\alpha\|\{111\}\delta$	$\{10\bar{1}7\}\alpha\|\{111\}\delta$	$\{10\bar{1}1\}\alpha\|\{110\}\delta$	Not applicable
Zr-1.0Nb	1	6	9	9
Zr-2.5Nb	1	9	9	6
Zircaloy-2	3	13	1	8
Zircaloy-4	0	12	6	7

CONCLUSIONS

The formation of hydrides in Zr-Nb alloys was evaluated by EBSP measurement and the orientation and the habit planes were compared with those in Zircaloys in this study. The α-matrix grains in Zr alloys have normal direction textures of (0001) basal planes. β-phase grains in Zr-1.0Nb and Zr-2.5Nb have no orientation. δ-hydride grains in four specimens have normal direction textures of {001} basal planes. In this study, $\{10\bar{1}7\}_{\alpha}||\{111\}_{\delta}$ is more common hydride habit plane relationship than $(0001)_{\alpha}||\{111\}_{\delta}$. On the other hand, the third relationship $\{10\bar{1}1\}_{\alpha}||\{110\}_{\delta}$ was hardly observed in Zircaloy-2, but often observed in Zr-1.0Nb, Zr-2.5Nb and Zircaloy-4.

REFERENCES

[1] M. Koike, S. Onose, K. Nagamatsu, and M. Kawajiri, Hydrogen pickup and degradation of heat-treated Zr-2.5 wt% Nb pressure tube, *JSME International Journal, Series B: Fluids and Thermal Engineering*, **36**(3), 464-70 (1993).

[2] S. Yamanaka, M. Miyake, and M. Katura, Study on the hydrogen solubility in zirconium alloys, *J. Nucl. Mater.*, **247**, 315-21, (1997).

[3] C.E. Coleman and J.F.R. Ambler, Solubility of hydrogen isotopes in stressed hydride-forming metals, *Script Metallurgica*, **17**(1), 77-82 (1983).

[4] J.J. Kearns, Terminal solubility and partitioning of hydrogen in the alpha phase of zirconium, Zircaloy-2 and Zircaloy-4, *J. Nucl. Mater.*, **22,** 292-303 (1967).

[5] C.D. Cann and A. Aterns, A metallographic study of the terminal solubility of hydrogen in zirconium at low hydrogen concentrations, *J. Nucl. Mater.*, **88,** .42-50 (1980).

[6] C.E. Coleman and D. Hardie, Hydrogen embrittlement of α-zirconium. A review., *J. Less-Common Met.*, **11**(3), 168-85 (1966).

[7] D. Srivastava, G.K. Dey, and S. Banerjee, Evolution of microstructure during fabrication of Zr-2.5 wt pct Nb alloy pressure tubes, *Metall. Mater. Trans.*, A **26**A, 2707-18 (1995).

[8] N. Nagai, T. Kakuma, and K. Fujita, Texture control of Zircaloy tubing during tube reduction, *Zirconium in the Nuclear Industry*, ASTM STP **754**, 26-38 (1982).

[9] K. Une and S. Ishimoto, EBSP measurements of hydrogenated Zircaloy-2 claddings with stress-relieved and recrystallized annealing conditions, *J. Nucl. Mater.*, **357**, 147-55 (2006).

[10] J.H. Root, W.M. Small, D. Khatamian, and O.T. Woo, Kinetics of the δ to γ zirconium hydride transformation in Zr-2.5Nb, *Acta Mater.*, **51**, 2041-53 (2003).

[11] K. Vaibhaw, S.V.R. Rao, S.K. Jha, N. Saibaba, and R.N. Jayaraj, Texture and hydride orientation relationship of Zircaloy-4 fuel clad tube during fabrication for pressurized heavy water reactors, *J. Nucl. Mater.*, article in press (2007).

[12] K. Une, K. Nogita, S. Ishimoto, and K. Ogata, Crystallography of zirconium hydrides in recrystallized zircaloy-2 fuel cladding by electron backscatter diffraction, *J. Nucl. Sci. Technol.*, **41**, 731-40 (2004).

[13] H.M. Chung, R.S. Daum, J.M. Hiller, and M.C. Billone, Characteristics of hydride precipitation and reorientation in spent-fuel cladding, *Zirconium in the Nuclear Industry*, ASTM STP **1423**, 561-582 (2002).

[14] V. Perovic and G.C. Weatherly, Hydride precipitation in α /β -zirconium alloys, *Acta Metall.*, **31,** 1381-91 (1983).

[15] V. Perovic and G.C. Weatherly, The nucleation of hydrides in a zirconium-2.5 wt% niobium alloy, *J. Nucl. Mater.*, **126**, 160-9 (1984).

[16] Y.S. Kim, Y. Perlovich, M. Isaenkova, S.S. Kim, and Y.M. Cheong, Precipitation of reoriented hydrides and textural change of α-zirconium grains during delayed hydride cracking of Zr-2.5%Nb pressure tube, *J. Nucl. Mater.*, **297,** 292-302 (2001).

[17] S. Neogy, D. Srivastava, R. Tewari, R.N. Singh, G.K. Dey, and S. Banerjee, Microstructural study of hydride formation in Zr–1Nb alloy, *J. Nucl. Mater.*, **322,** 195-203 (2003).

FEM STUDY OF DELAYED HYDRIDE CRACKING IN ZIRCONIUM ALLOY FUEL CLADDING

Masayoshi Uno[1], Masato Ito[1], Hiroaki Muta[1], Ken Kurosaki[1], Shinsuke Yamanaka[1]
[1]Graduate School of Engineering, Osaka University, Suita, Osaka 565-0871, Japan
Keizo Ogata[2]
[2]Japan Nuclear Energy Safety Organization, Minato-ku, Tokyo 105-0001, Japan

ABSTRACT

Thermal-displacement coupled analysis was performed for models of the BWR cladding using the finite element method in order to gain understanding of DHC failure mechanism. The negative hydrostatic pressure spread in a circular pattern around the crack tip. The hydrostatic pressure increased with the internal pressure. The hydrostatic pressure and the stress-focusing area around crack tip increased with the crack length. Using the calculated distribution of the hydrostatic pressure, transient hydrogen diffusion analysis, considering the three effects of hydrogen concentration, temperature, and stress, was performed. The hydrogen distribution after the internal pressure loading was noticeably different from the initial state and a large amount of hydrogen piled up near the crack tip. The hydrogen diffused into the vicinity of the crack tip and the amount of piled hydrogen increased with increasing the internal pressure. With increasing the crack length, the more amount of hydrogen piled up around the crack tip and the hydrogen-focusing area significantly spread.

INTRODUCTION

Breakage failures of the claddings in a different form from those observed so far and considerably lower power levels for the failure are reported as a result of power ramp tests of high burnup Boiling Water Reactor (BWR) fuels in Japan [1, 2]. It is considered that the failure process started with an axial split by cracking of radial hydrides which were formed during the power ramp test, followed by propagation caused by step-by-step cracking of hydrides at a crack tip [1]. This process is generally called as delayed hydride cracking (DHC) [3-6], which is based on the diffusion of hydrogen to the crack tip with higher tensile stress, followed by hydride nucleation, its growth, and fracture of the hydride. Since the crack velocity depends on time required for hydrogen to accumulate enough for precipitation and for growing of hydride to the critical size hydrogen diffusion to the crack tip from the surroundings is one of the most important factors for the failure process. However, the hydrogen behavior in the fuel cladding during reactor operation is affected by the conditions varying from hour to hour, and is quite complex. And it is difficult to evaluate the mechanism of the DHC process only from the experimental work. Therefore it is need to study parametrically by computer simulations in order to gain the understanding of the failure mechanism.

In the present study, transient hydrogen diffusion analysis considering three effects of hydrogen concentration, temperature, and stress was performed using finite element method (FEM) under the conditions of power ramp tests. At first the temperature and stress distributions in a modeled cladding with cracks were calculated. Then, the hydrogen diffusion analysis was performed using the calculated distributions as the initial condition. The variation of hydrogen content with the radial distance from the crack tip was discussed as a function of internal pressure or crack length.

CALCULATION PROCEDURE

Fig. 1 shows the two-dimensional model and boundary conditions used in the calculations. The fuel cladding with outside cracks which have a semicircular tip with 1 μm diameter was simulated. Since the stress gradient near the crack tip would be quite sharp the mesh was finely divided as it approached to the tip. The calculation was done for the second-order isoparametric elements using ABAQUS ver. 6.5. Thermophysical properties of irradiated Zry-2 [7] were used for the calculation. Hydrogen solubility was obtained by the extrapolation of the high-temperature literature values [8].

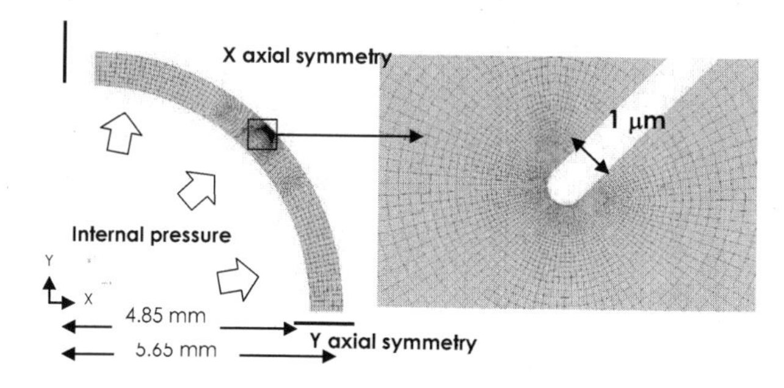

Fig. 1 Computational system and divided meshes around outside cracks.

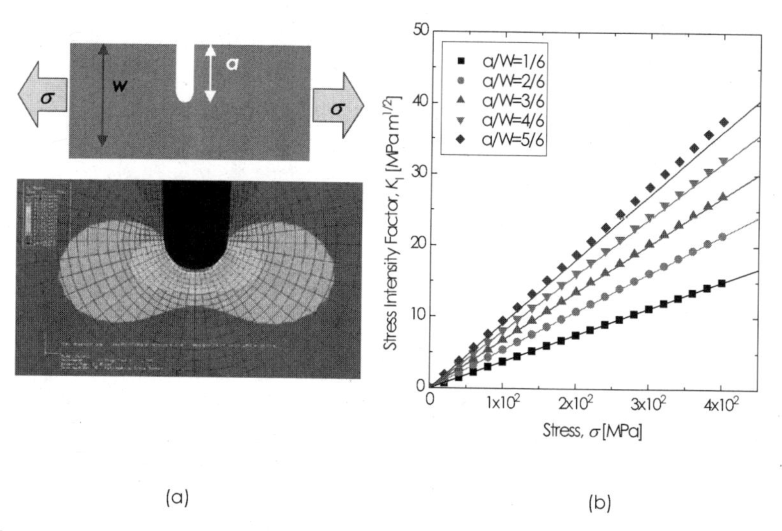

Fig. 2 (a)flat plate model with a crack and distribution of the Mises stress near the crack tip, (b) calculated stress intensity factor together with the analytical solutions.

In order to confirm the validity of the modeling, stress intensity factor for a crack of flat plate sample was calculated and compared with analytical solution[9]. Fig. 2(a) and (b) show Mises stress distribution near the crack and the calculated stress intensity factors together with the analytical solutions, respectively. The calculated values are in good accordance with the analytical solutions. Thus, the modeling of crack and mesh division were considered to be appropriate.

The diffusion is assumed to be driven by the gradient of chemical potential of hydrogen which is defined as partial differentiation of free energy. Gibbs free energy is defined as follows:

$$G \equiv E - TS + PV \tag{1}$$

Here, E, T, S, P, and V are internal energy, absolute temperature, entropy, pressure and volume, respectively. E and S of metal-hydrogen system are evaluated by the classical theory as follows:

$$E = E_{\mathrm{METAL}} + C_{\mathrm{H}} E_{\mathrm{H}} + C_{\mathrm{H}} \cdot \Delta E_{\mathrm{M-Hbond}} \tag{2}$$

$$S = S_{\mathrm{METAL}}^{\mathrm{vib}} + C_{\mathrm{H}} S_{\mathrm{H}}^{\mathrm{vib}} + C_{\mathrm{H}} \cdot \Delta S_{\mathrm{M-Hbond}} + R \ln \frac{N!}{n!(N-n)!} \tag{3}$$

Here, E_{METAL} and E_{H} are molar energy of pure metal and hydrogen, $S_{\mathrm{METAL}}^{\mathrm{vib}}$, and $S_{\mathrm{H}}^{\mathrm{vib}}$ are molar entropy of metal and hydrogen vibration, C_{H} is hydrogen content, $\Delta E_{\mathrm{M-Hbond}}$, and $\Delta S_{\mathrm{M-Hbond}}$ are change in the molar energy and the entropy due to creation of M-H bonds, R is the gas constant, N is number of interstitial sites for hydrogen in lattice, and n is number of occupied sites in N.

The present calculations were performed on the assumption that Sieverts' law holds. In other words, $\Delta E_{\mathrm{M-Hbond}}$ and $\Delta S_{\mathrm{M-Hbond}}$ are assumed to be independent of hydrogen content. In that case, we can get the chemical potential of hydrogen using above parameters as follows:

$$\begin{aligned}
\mu_{\mathrm{H}} &= \left(\frac{\partial G}{\partial C_{\mathrm{H}}}\right) \\
&= (E_{\mathrm{H}} + \Delta E_{\mathrm{M-Hbond}}) - T \cdot (S_{\mathrm{H}}^{\mathrm{vib}} + \Delta S_{\mathrm{M-Hbond}}) + \mathrm{R}T \ln C_{\mathrm{H}} + P\overline{V}_{\mathrm{H}} \\
&= \mathrm{R}T \ln Ks + \mathrm{R}T \ln C_{\mathrm{H}} + P\overline{V}_{\mathrm{H}} \\
&= \mathrm{R}T \ln\left(\frac{p^{1/2}}{C_{\mathrm{H}}} \cdot \frac{1}{p_0^{1/2}}\right) + \mathrm{R}T \ln C_{\mathrm{H}} + P\overline{V}_{\mathrm{H}} \\
&= \mu_0 + \mathrm{R}T \ln(C_{\mathrm{H}} / s) + P\overline{V}_{\mathrm{H}}
\end{aligned} \tag{4}$$

Here, $\overline{V}_{\mathrm{H}}$ is partial molar volume of hydrogen, Ks is Sieverts' constant, p is equilibrium hydrogen pressure at C_{H}, p_0 is standard hydrogen pressure, μ_0 is chemical potential of hydrogen at the standard condition, and s is solubility of hydrogen.

The diffusion equation for hydrogen is obtained by extending Fick's equation to allow for nonuniform solubility of the hydrogen in material. Using the Fick's equation and the equation (4), hydrogen diffusion flux J is calculated as follows:

$$
\begin{aligned}
J &= -\frac{C_{\mathrm{H}} D}{\mathrm{R}T} \cdot \frac{\partial \mu_{\mathrm{H}}}{\partial x} \\
&= -\frac{C_{\mathrm{H}} D}{\mathrm{R}T} \cdot \left(\frac{\mathrm{R}T}{C_{\mathrm{H}}} \cdot \frac{\partial C_{\mathrm{H}}}{\partial x} + \mathrm{R} \ln(C_{\mathrm{H}} / s) \cdot \frac{\partial T}{\partial x} + \overline{V}_{\mathrm{H}} \cdot \frac{\partial P}{\partial x} \right) \\
&= -D \left(\frac{\partial C_{\mathrm{H}}}{\partial x} + \frac{C_{\mathrm{H}} \cdot \ln(C_{\mathrm{H}} / s)}{T} \cdot \frac{\partial T}{\partial x} + \frac{C_{\mathrm{H}} \cdot \overline{V}_{\mathrm{H}}}{\mathrm{R}T} \cdot \frac{\partial P}{\partial x} \right) \\
&= -D \left(\frac{\partial C_{\mathrm{H}}}{\partial x} + k_1 \cdot \frac{\partial T}{\partial x} + k_2 \cdot \frac{\partial P}{\partial x} \right)
\end{aligned}
\tag{5}
$$

Here, D is hydrogen diffusion coefficient and k is a proportional coefficient. In this equation, the first, second, and last terms are the contribution to the diffusion due to gradients of hydrogen content, temperature, and hydrostatic pressure, respectively.

k_1 and k_2 can be calculated based on the temperature dependence of hydrogen solubility and the partial molar volume of hydrogen, respectively. The hydrogen distribution due to temperature gradient is often measured for several materials under a steady state condition without stress. In that case, the Fick's equation can be expressed as follows:

$$
\frac{\partial C_{\mathrm{H}}}{\partial x} = -Q^* \cdot \frac{\partial T}{\partial x} \tag{6}
$$

Here, Q^* is heat of hydrogen transport. In the present study, literature data of Q^*=25.12 kJ/mol [10] is employed. Therefore, the equation (5) is rewritten as follows:

$$
J = -D \left(\frac{\partial C_{\mathrm{H}}}{\partial x} + Q^* \cdot \frac{\partial T}{\partial x} + k_2 \cdot \frac{\partial P}{\partial x} \right) \tag{7}
$$

Ideally speaking, the coefficient k_2 should be also measured because the heat of hydrogen transport Q^* does not always equal to the calculated coefficient k_1. Since terminal solid solubility of hydrogen in zirconium alloys at around 600 K is quite low, there is insufficient experimental information on the hydrogen partial molar volume in Zr. The hydrogen partial molar volume for hcp metals is commonly about $3.0\,\mathrm{nm}^3$/(H/M) [10], which is adopted for calculating coefficient k_2.

Based on the information of power ramp tests, inside and outside temperatures of the cladding were fixed to be 613 K and 573 K, respectively. Although the internal pressure of the BWR cladding during the normal operation is ranging from 1 to 4 MPa the calculation was done for the internal pressure ranging from the 5 to 50 MPa because further stress is assumed to be locally applied from inner surface by the PCMI during serious ramp tests. In order to discuss the behavior of crack propagation due to the DHC, the calculations were also performed for different crack lengths.

RESULTS AND DISCUSSION

Stress and Strain Distribution of Fuel Cladding with Crack

Fig. 3 shows calculation results of hydrostatic pressure for a model without crack; (a) distribution of the pressure, (b) variation of the pressure with the radial distance. As shown in Fig. 3(b), the hydrostatic pressure linearly decreases with increasing the distance from the inner surface even in the case of zero internal pressure due to the temperature gradient. The hydrostatic pressure has negative values except for that inside from middle line of the cladding at 0 MPa internal pressure and thus, there are usually a tensile stress and its absolute value increases with the internal pressure. Since the tensile stress also increases with the distance from the inner surface hydrogen in the cladding diffuses towards outside due to the stress gradient as well as the temperature gradient.

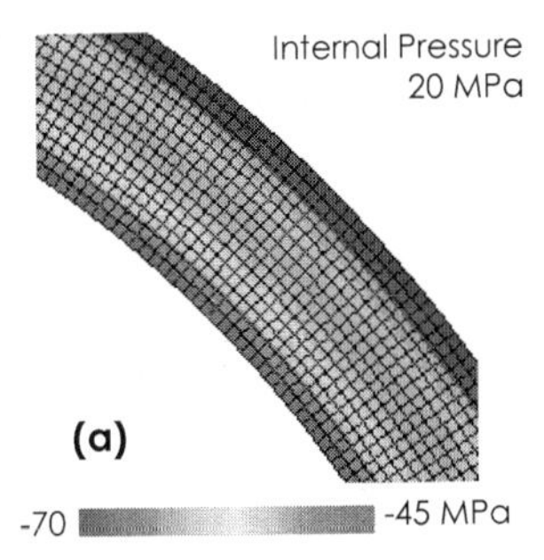

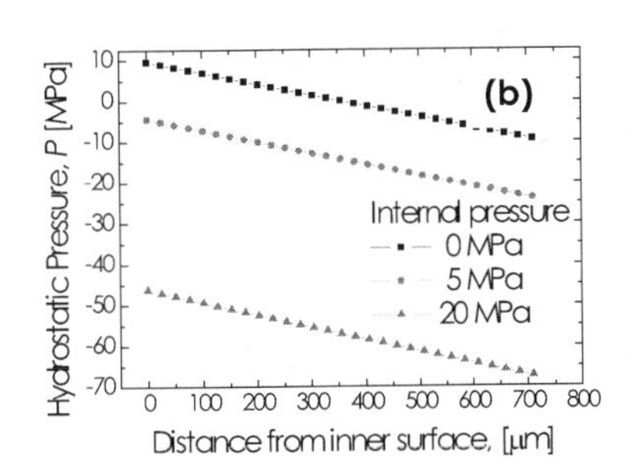

Fig. 3 Calculation results of hydrostatic pressure for a model without crack, (a) distribution of the pressure, (b) variation with the distance from the inner surface.

Fig. 4 shows the distribution of radial stress near the tips of different crack lengths where plus stress corresponds to tensile stress. The internal pressure of all the case equals to 5 MPa. The stress value and stress-focusing area near crack tip increase with crack length, although its distribution profile is not changed so much by the crack length.

Fig. 5 shows the distribution of circumferential stress for the same case where also the plus values correspond to tensile stress. It is found from this figure that there is quite higher tensile stress in circumferential direction around the crack tip than the radial direction. Similar to the case of the radial stress, the stress value and stress-focusing area near crack tip increase with the crack length.

The distribution of hydrostatic pressure, that is mean value of the above two stresses is shown in Fig. 6, where minus values correspond to tensile stress. The negative hydrostatic pressure spreads in a circular pattern around the crack tip. In addition, vicinity of crack surface slightly away from the tip has low positive hydrostatic pressure. This suggest that negative hydrostatic pressure field near crack tip, in which hydrogen piles up, shifts to the positive field when the crack would proceed. Therefore, hydrogen is considered to easily diffuse from the crack surface to the tip.

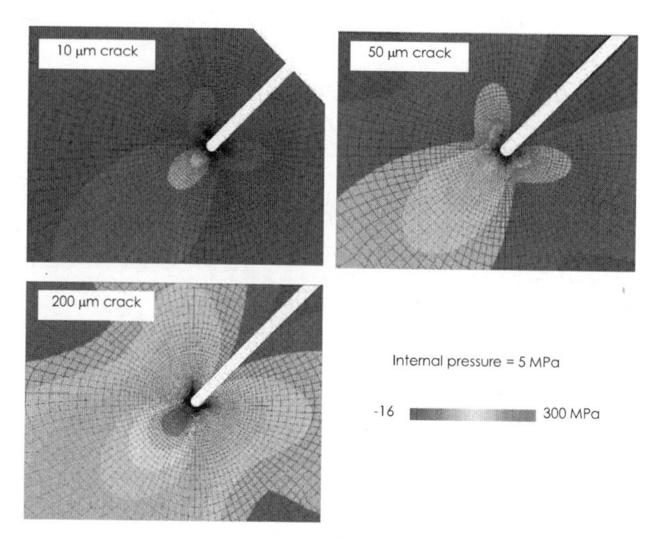

Fig. 4 Distribution of radial stress near crack tips with different lengths.

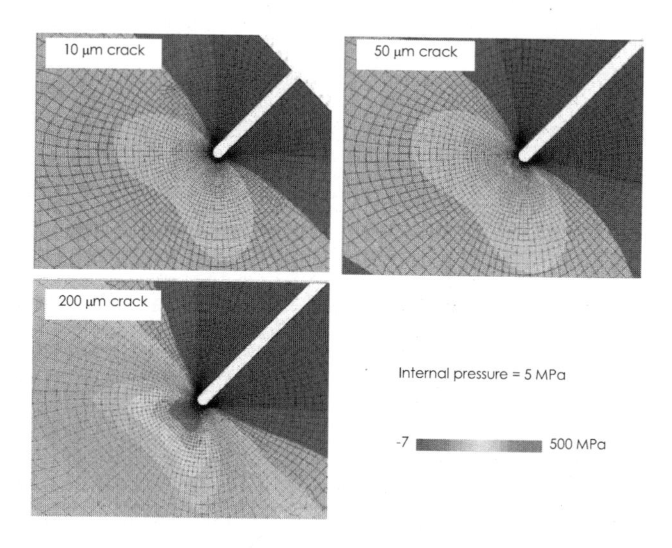

Fig. 5 Distribution of circumferential stress near crack tips with different lengths.

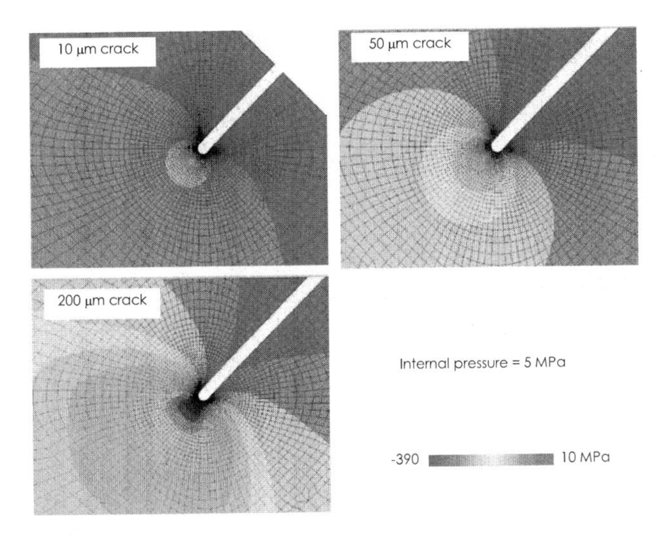

Fig. 6 Distribution of hydrostatic pressure near crack tips with different lengths.

Fig. 7 shows the variation of hydrostatic pressure with the radial distance from the tips of 50 μm crack with different internal pressures. Fig. 7(b) is a detailed drawing near the crack tip. The hydrostatic pressure increases with the internal pressure. In the case of 50 μm crack, this trend is pronounced within about 100 μm distant from the crack tip. A maximum point of the hydrostatic pressure is a few micro-meters distant from the tip. This is because the cladding in the close vicinity of the crack tip is plastically-deformed and highly focused stress is relaxed as shown in Fig. 7(b).

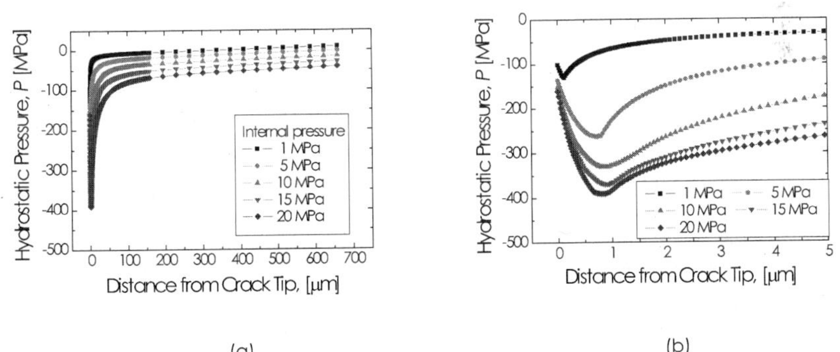

Fig. 7 Variation of hydrostatic pressure with distance from 50 μm crack tips.

Fig. 8 shows the variation of hydrostatic pressure with the distance from the crack tips with different lengths under the internal pressure of 20 MPa. Fig. 8(b) is also a detailed drawing near the crack tip. The hydrostatic pressure and stress-focusing area around crack tip increase with increasing the crack length. In the case of 200μm crack, the hydrostatic pressure shifts from negative to positive around 400 μm form the crack tip, hence hydrostatic pressure gradient is widely spreading across the cladding. These results are considered to profoundly affect hydrogen behaviors.

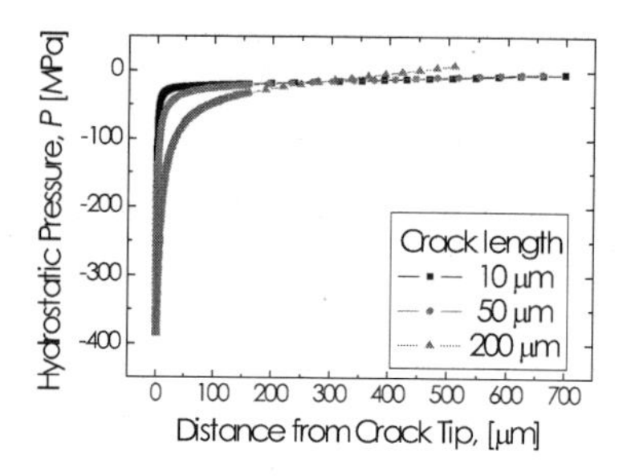

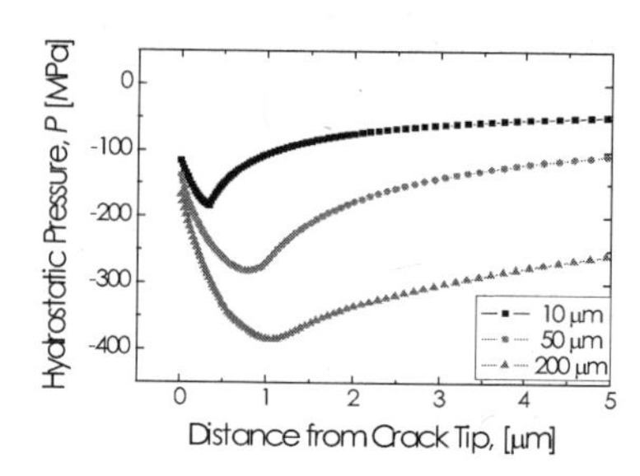

(a) (b)

Fig. 8 Variation of hydrostatic pressure with the distance from crack tips at 20 MPa.

Hydrogen Diffusion Behavior under Gradients of Hydrogen, Temperature, and Stress

Hydrogen diffusion only by the temperature gradient was analyzed and its result was used as the initial hydrogen distribution for the evaluation of hydrogen distribution by the stress. Since a zirconium liner on inner surface of the BWR cladding contains a larger amount of hydrogen than Zry-2 during reactor operation, the zirconium liner was regarded as a source of hydrogen and the hydrogen diffusion analysis was performed on the boundary condition that hydrogen content of inner surface is fixed to be 100 ppm. Fig. 9(a) shows the distribution of hydrogen content near the tip of 50 μm crack at the initial state. The hydrogen content in outer region is higher than that in inner region because the hydrogen solubility of zirconium decreases with increasing temperature. In the present boundary condition, mean value of hydrogen content in the cladding approximately equals to 180 ppm. Fig. 9(b) shows the distribution of hydrogen content with internal pressure of 5 MPa at equilibrium state after the internal pressure loading. It is found from this figure that the hydrogen distribution after the internal pressure loading is noticeably different from the initial state and a large amount of hydrogen piles up near the crack tip.

This is also confirmed from Fig. 10 that shows the variation of hydrogen content with the radial distance from the crack tip. Similar to the case of the stress distributions, the hydrogen piles up within about 100 μm distant from the crack tip and a maximum point of the hydrogen content is a few micro-meters distant from the tip. Comparing Fig. 9(a) and (b), it is also found that hydrogen content at vicinity of crack surface, which is slightly away from the tip, reduces by the internal pressure loading. These trends correspond to the stress distribution.

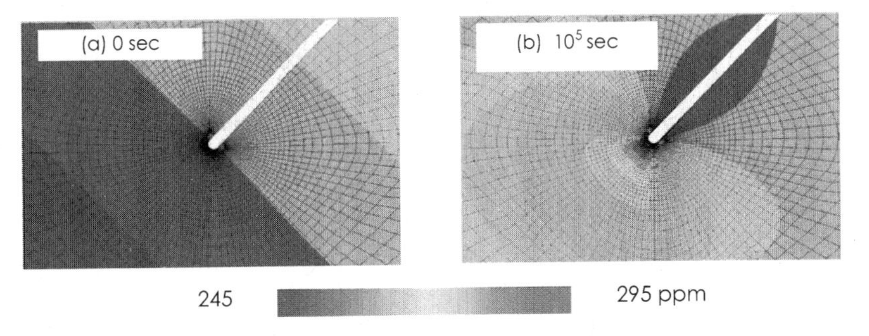

Fig. 9 Distribution of hydrogen content near tip of 50 μm crack with internal pressure of 5 MPa, (a) the Initial state, (b) equilibrium state after the internal pressure loading.

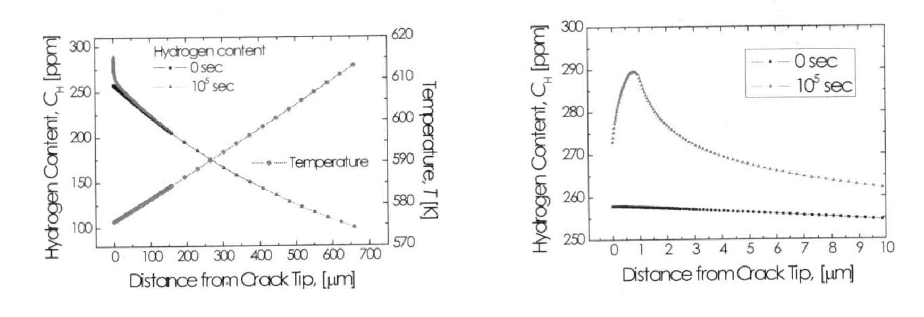

Fig. 10 Radial directional distributions of hydrogen content from tips of 50 μm crack with internal pressure of 5 MPa.

Fig. 11 shows the changing rate of radial directional distributions of hydrogen content from the crack tips with different internal pressure, in which the crack length equals to 50 μm. Although the distribution profile and the maximum point of the hydrogen content are not so changed by the internal pressure, the amount of hydrogen that is piled up near the crack tip increases with increasing the internal pressure. This implies that the PCMI during the power ramp tests may encourage the DHC breakage failure.

Fig. 12 shows the changing rate of radial directional distributions of hydrogen content from the crack tips with different crack lengths. With increasing the crack length, the much amount of hydrogen piles up around the crack tip and the hydrogen-focusing area significantly spreads. Therefore, it is considered that the crack easily penetrates the cladding if the length reaches a certain level.

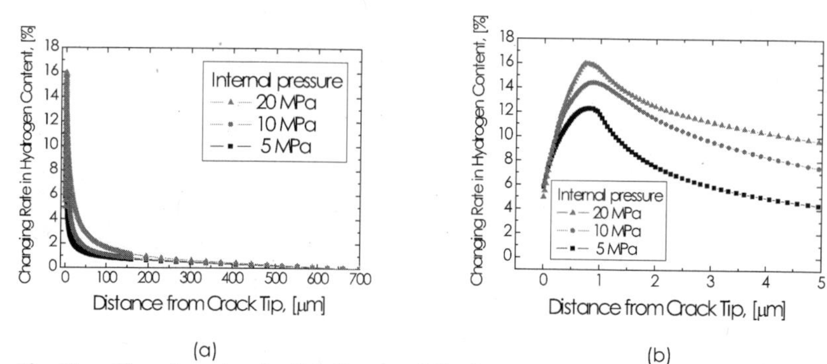

Fig. 11 Changing rate of radial directional distributions of hydrogen content from crack tips with different internal pressure (Crack lengths equal to 50 μm).

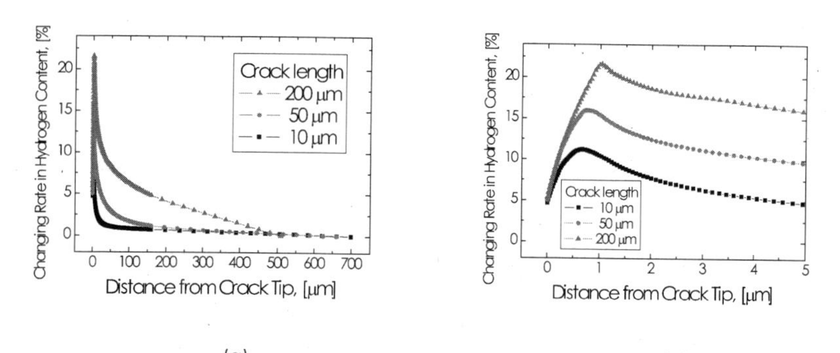

Fig. 12 Changing rate in radial directional distributions of hydrogen content from crack tips with different lengths (Internal pressure equals to 20 MPa).

As mentioned above the internal pressure is larger or the crack is longer the changing rate of hydrogen becomes large faster, that is the velocity of hydrogen pile up is larger. The present study successfully provided tangible information on hydrogen behavior during the power ramp test and the qualitative effects of the several conditions were revealed. However, there is marked variability in the literature data of hydrogen diffusion coefficient. In addition, the diffusion coefficient and solubility of hydrogen would change due to the external stress. Therefore, the sensitivity analysis will be important in the next step for the investigation. It is also important to estimate the velocity of the hydrogen pile up as a future work to mechanistic study of DHC.

CONCLUSION

The thermal-displacement coupled analysis was performed for the models of the BWR cladding using the finite element method. The negative hydrostatic pressure spread in a circular pattern around the crack tip and the hydrostatic pressure increased with the internal pressure. The hydrostatic pressure and the stress-focusing area around crack tip increased with the crack length. Using the distribution of the hydrostatic pressure, the transient hydrogen diffusion analysis, considering the three effects of hydrogen concentration, temperature, and stress, was performed. The hydrogen distribution after the internal pressure loading was noticeably different from the initial state and a large amount of hydrogen piled up near the crack tip. The hydrogen diffused into the vicinity of the crack tip and the amount of piled hydrogen increased with increasing the internal pressure. With increasing the crack length, the more amount of hydrogen piled up around the crack tip and the hydrogen-focusing area significantly spread.

For future work on crack propagation due to the DHC, it is considered that the time for hydrogen to pile up near the crack tip must be clarified.

REFERENCES

[1] S. Shimada, E. Etoh, H. Hayashi, Y. Tukuta, A metallographic and fractographic study of outside-in cracking caused by power ramp tests, Journal of Nuclear Materials, 327 (2004) 97-113.

[2] M. Yamawaki, K. Konashi, S. Shimada, Journal of the Atomic Energy Society of Japan, 46 (2004) 457.(in Japanese)

[3] C.E. Coleman, S. Sagat, G.K. Shek, D.B. Graham, M.A. Durand, Locating a leaking crack by safe stimulation, International Journal of Pressure Vessels and Piping, 43 (1990) 187-204.

[4] S.Q. Shi, G. K. Shek and M. P. Puls, Hydrogen concentration limit and critical temperatures for delayed hydride cracking in zirconium alloys, Journal of Nuclear Materials, 218 (1995) 189-201.

[5] Y.S. Kim, S.B. Ahn, Y.M. Cheong, ,Precipitation of crack tip hydrides in zirconium alloys, Journal of Alloys and Compounds, 429 (2007) 221-226.

[6] R.N. Singh, S. Roychowdhury, V.P. Sinha, T.K. Sinha, P.K. De, S. Banerjee, Delayed hydride cracking in Zr–2.5Nb pressure tube material: influence of fabrication routes, Materials Science and Engineering A, 374 (2004) 342-350.

[7] SCDAP/RELAP5/MOD3.1 Code 4. MATPRO, INEL Report NUREG/CR-6150 EGG-2720, Idaho Falls, (1993).

[8] S. Yamanaka, M. Miyake, M. Katsura, Study on the hydrogen solubility in zirconium alloys, Journal of Nuclear Materials, 247 (1997) 315.

[9] T.L. Anderson, Fracture Mechanics (2nd Edition), CRC Press LLC, (2000).

[10] A. Sawatzky, Hydrogen in zircaloy-2: Its distribution and heat of transport, Journal of Nuclear Materials, 2 (1960) 321.

[11] Y. Fukai, The Metal-Hydrogen System: Basic Bulk Properties, Springer Series in Materials Science 21 (1993).

THE EFFECT OF MANGANESE STOICHIOMETRY ON THE CURIE TEMPERATURE OF $La_{0.67}Ca_{0.26}Sr_{0.07}Mn_{1+x}O_3$ USED IN MAGNETIC REFRIGERATION.

I. Biering, M. Menon, and N. Pryds

Fuel Cells and Solid State Chemistry Division, Risø National Laboratory for Sustainable Energy, Technical University of Denmark.

DK4000 Roskilde, Denmark

ABSTRACT

Polycrystalline samples of $La_{0.67}Ca_{0.26}Sr_{0.07}Mn_{1+x}O_3$ (x = -0.20, -0.15, -0.10, -0.05, -0.033, -0.017, 0.00, 0.05, 0.10, 0.15, 0.20) have been prepared using the glycine-nitrate combustion synthesis. The samples were characterized by x-ray diffraction (XRD) and the Curie temperature determined by differential scanning calorimetry (DSC). The Curie temperatures were found to decrease rapidly when the amount of manganese decreases below the stochiometric value (x=0.00). For x>0.00 there was no apparent effect on the Curie temperature. The reason for the variation in the Curie temperature with the composition is discussed in the text.

INTRODUCTION

Magnetic refrigeration near room temperature is a promising alternative technology to the conventional vapour compression technique. In a magnetic refrigerator the magnetocaloric effect is utilized in a heat pump cycle to achieve a cooling effect[1]. The magnetocaloric effect is largest in the vicinity of the Curie temperature, T_c. In order to exploit fully the magnetocaloric effect for near room temperature refrigeration, T_c should be close to the operating temperature of the refrigerator[2]. Manganese perovskites such as $(La,Ca,Sr)MnO_3$ (ABX_3) are among the candidate materials for room temperature magnetic refrigeration due to their moderate magnetocaloric effect and tuneable Curie temperature[1]. Earlier studies have shown that the manganese stoichiometry of $La_{0.67}Ca_{0.33}Mn_{1-x}O_3$ has a profound effect on the Curie temperature of this material[3]. The present work is part of a large afford to study the relationship between powder processing, sintering behaviour and the magnetocaloric effect. The aim of this work was to investigate the effect of manganese stoichiometry on the Curie temperature of $La_{0.67}Ca_{0.26}Sr_{0.07}Mn_{1+x}O_3$.

EXPERIMENTAL DETAILS

Eleven samples of polycrystalline $La_{0.67}Ca_{0.26}Sr_{0.07}Mn_{1+x}O_3$ (x = -0.20, -0.15, -0.10, -0.05, -0.033, -0.017, 0.00, 0.05, 0.10, 0.15, 0.20) were prepared by the glycine-nitrate combustion synthesis[4]. Aqueous solutions of $La(NO_3)_3{\bullet}4H_2O$, $Ca(NO_3)_2{\bullet}4H_2O$, $Sr(NO_3)_2{\bullet}H_2O$ and $Mn(NO_3)_2{\bullet}4H_2O$, were calibrated and mixed corresponding to the desired compositions. After addition of glycine (glycine/nitrat = 0.588) the solutions were heated until the water had evaporated and spontaneous combustion occurred resulting in fine grained (La,Ca, Sr)Mn_xO_3 powders. The powders were calcined at 700°C for 4 hours and then pressed into pellets which were sintered in air at 1200°C for 4 hours.

Structural characterization was performed by x-ray diffraction (XRD) at room temperature using a Bruker D8 Advance diffractometer. $\theta - 2\theta$ scans were performed on the pellets in the range of 20 – 80°. The radiation used was Cu-Kα, scanning rate 5 s/step and step size 0.02°. Curie temperatures were determined by Differential Scanning Calorimetry (DSC) using a Netzsch DSC 200F3 Maia. The temperature range of the DSC measurement were from -50°C to 50°C with a heating rate of 1K/minute. The Curie temperatures were defined as the maximum peak height of the phase transition peaks.

RESULTS AND DISCUSSION

Figure 1 shows the XRD of compositions x = -0.20, 0.00 and 0.20. The XRD patterns showed that the main constituent of all the prepared samples was the orthorhombic perovskite structure, $La_{0.67}Ca_{0.26}Sr_{0.07}Mn_{1+x}O_3$. Samples with excess manganese (x=0.00, 0.05, 0.15 and 0.20) were essentially single phase whereas samples with x<1 showed an increasing amount of secondary phases which have not been completely identified yet. Patterns for x = 0.05, 0.10 and 0.15 were similar to that of x = 0.20, and patterns for x = -0.017, -0.033, -0.05, -0.10, -0.15 and -0.20 showed an increasing amount of secondary phases with a significant increase going from x = -0.15 to x = -0.20.

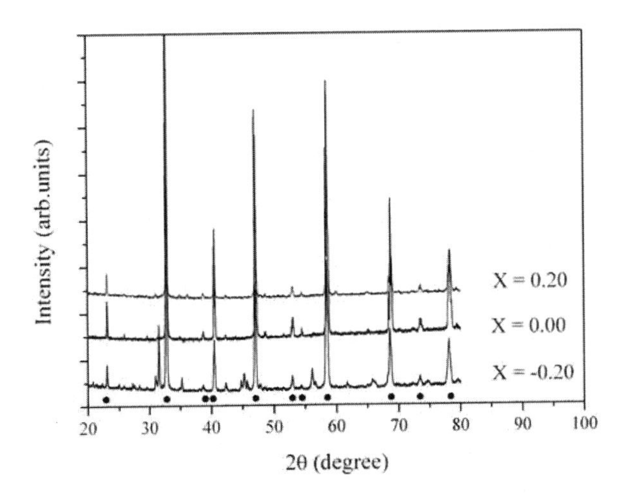

Figure 1. X-ray diffraction patterns for $La_{0.67}Ca_{0.26}Sr_{0.07}Mn_{1+x}O_3$, x=0.20, 0.00 and -0.20. The dots indicate the expected peaks for the orthorhombic perovskite structure.

In Figure 2 the lattice parameters and the unit cell volume is plotted as a function of the manganese stoichiometry. The figure shows that the material undergoes a change in the lattice parameters and the unit cell volume from the under-stochiometric compositions (x<0) to the over-stochiometric compositions (x>0). The lattice parameters seem to decrease when approaching the stoichiometric composition and then keep relatively constant for values above the stochiometric composition. The fact that the lattice parameters increases when increasing the Mn deficiency (x<0) up to x = -0.10 implies that the $La_{0.67}Ca_{0.26}Sr_{0.07}Mn_{1+x}O_3$ composition can accommodate up to 10% Mn deficiency. Chen et.al[3] also observed the same expansion in the lattice parameter for compositions below the stochiometric value for $La_{0.67}Ca_{0.33}Mn_{1-x}O_3$ ($0.00 \leq x \leq 0.06$). The fact that lattice parameters and unit cell volume did not change for compositions with over-stoichiometric manganese (x>0) implies that the $La_{0.67}Ca_{0.26}Sr_{0.07}Mn_{1+x}O_3$ composition does not accommodate excess Mn in the unit cell.

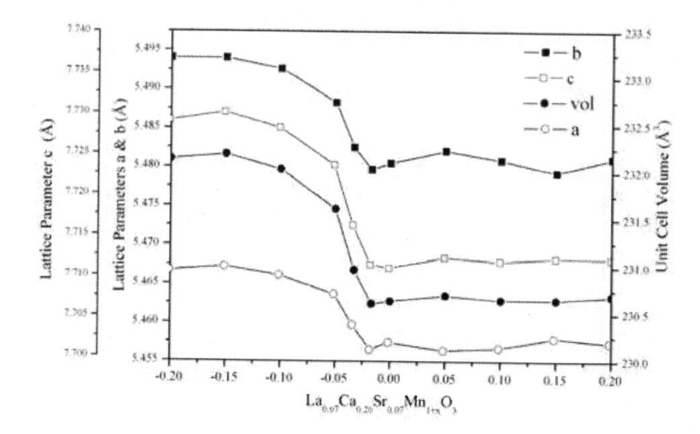

Figure 2. Lattice parameters and unit cell volume as a function of manganese stoichiometry.

The Curie temperature, T_c, as determined by DSC is plotted in Figure 3 as a function of Mn stoichiometry. Stoichiometric composition (x = 0) exhibited a T_c of 20°C which dropped to -5°C for the x = -0.05 composition. Below 5% Mn deficiency no magnetic phase transition was detected in the measured temperature range. A similar dependence of the T_c on the manganese stoichiometry has been reported by Chen et.al[3] for $La_{0.67}Ca_{0.33}Mn_{1-x}O_3$ ($0.00 \leq x \leq 0.06$).

The T_c of manganese based perovskites is affecte by the Mn–O–Mn bond angle. Therefore, distortions in the MnO_6 octahedra in the perovskite structure will lead to changes in the T_c of manganese based perovskites[5]. As the unit cell of $La_{0.67}Ca_{0.26}Sr_{0.07}Mn_{1+x}O_3$ composition expands with increasing Mn deficiency (see Figure 2), the Mn–O distances probably increases, leading to smaller Mn–O–Mn angles which corresponds to weaker exchange interactions and lower T_c.

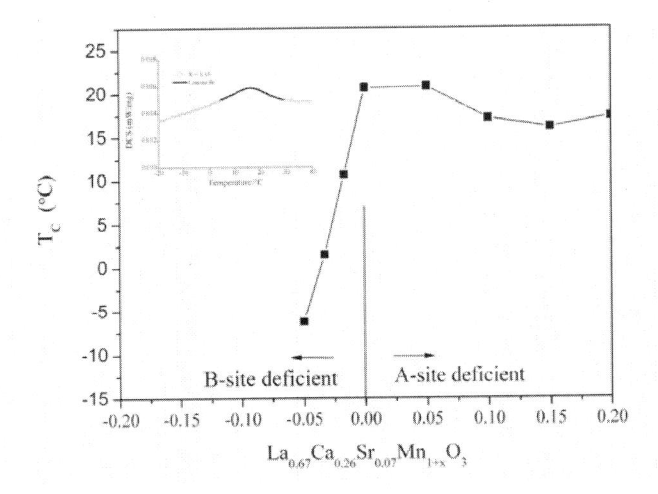

Figure 3. The Curie temperature as a function of the manganese stoichiometry. The Curie temperature was determined as the peak maximum of the DSC curve as illustrated by the graph in the upper left corner.

Since the lattice parameters does not change significantly for $La_{0.67}Ca_{0.26}Sr_{0.07}Mn_{1+x}O_3$ (x>0) compositions (see Figure 2), T_c is expected to remain unchanged. Nevertheless, the T_c decreases slightly for x>0 compositions as shown in Figure 3. In this discussion, we have assumed that the oxygen stoichiometry in $La_{0.67}Ca_{0.26}Sr_{0.07}Mn_{1+x}O_3$ is three. However, as the samples are sintered at 1200°C, the samples probably loose oxygen[6]. Since the samples after the sintering were cooled to room temperature rapidly, it is likely that the they are oxygen deficient. The oxygen deficiency can therefore also be a factor affecting the transition temperature due to the changes in the Mn–O–Mn angle. The effect of Mn and oxygen deficiency on the Mn–O–Mn angle and hence on the magnetic transition temperature are a subject for the future work.

CONCLUSIONS

The manganese stoichiometry has a significant influence on the Curie temperature of $La_{0.67}Ca_{0.26}Sr_{0.07}Mn_{1+x}O_3$. Manganese deficiency rapidly decreases the T_c. Mn excess, on the other hand shows little or no effect on the T_c. The measurements of the Curie temperature were found to be consistent with the XRD data. Finally, care about the exact composition of these materials must be taken when processing these materials for the use in magnetic refrigeration.

ACKNOWLEDGEMENTS

The authors would like to acknowledge the support of the Programme Commission on Energy and Environment (EnMi) (Contract No. 2104-06-0032) which is part of the Danish Council for Strategic Research.

REFERENCES

[1]A. R. Dinesen, S. Linderoth, and S. Mørup, Direct and indirect measurement of the magnetocaloric effect in $La_{0.67}Ca_{0.33}Mn_{1-x}O_{3\pm\delta}$ (x € [0;0.33]), *Journal of Physics: Condensed Matter*, **17**, 6257-6269 (2005).

[2]K. A. Geschneidner Jr, V. K. Pecharsky, and A. O. Tsokol, Recent developments in magnetocaloric materials, Reports on Progress in Physics, 68, 1479-1539 (2005).

[3]W. Chen, L. Y. Nie, X. Zhao, W. Zhong, G. D. Tang, A. J. Li, J. J. Hu, and Y. Tian, Effect of Mn-site vacancies on the magnetic entropy change and Curie temperature of $La_{0.67}Ca_{0.33}Mn_{1-x}O_3$ perovskite, *Solid State Communications*, **138**, 165-168 (2006).

[4]L. A. Chick, L. R. Pederson, G. D. Maupin, J. L. Bates, L. E. Thomas, and G. J. Exarhos, Glycine-nitrate combustion synthesis of oxide ceramic powders, *Materials Letters*, **10**, 6-12 (1990).

[5]W. Chen, W. Zhong, D. L. Hou, R. W. Gao, W. C. Feng, M. G. Zhu, and Y. W. Du, Preparation and magnetocaloric effect of self-doped $La_{0.8-x}Na_{0.2}\square_xMnO_{3+\delta}$, (□=vacancies) polycrystal, *Journal of Physics: Condensed Matter*, **14**, 11889-11896 (2002)

[6]F. W. Poulsen. Defect chemistry modelling of oxygen-stoichiometry, vacancy concentrations, and conductivity of $(La_{1-x}Sr_x)_yMnO_{3\pm\delta}$, *Solid State Ionics*, **129**, 145 – 162 (2000).

PREPARATION OF ELECTROCATALYTICALLY ACTIVE RuO_2/Ti ELECTRODES BY PECHINI METHOD

O. Kahvecioglu[a,*], S. Timur[a]
[a]: Istanbul Technical University, Department of Metallurgical and Materials Engineering Istanbul, Turkey

ABSTRACT

In this study, DSA® type porous ruthenium oxide coated titanium electrodes with different coating masses were prepared by Pechini method from a polymeric precursor at different decomposition temperatures. The influences of temperature on the morphology and the microstructure of RuO_2 formed on titanium were investigated by SEM and thin film-XRD. This work aimed to prepare electrocatalytically active electrodes showing high stability during oxygen evolution reaction (OER). Linear Sweep Voltammetry (LSV) tests were performed in 1 M H_2SO_4 after a 1-h-constant anodic polarization at 20 mA/cm^2. Tafel slopes (*b*) of 55-60 mV/dec were obtained. The electrodes prepared at lower temperatures show lower anode potentials than that of prepared at higher temperatures. Increasing of the coating mass showed an increase in the coating stability of the electrode. Similar electrocatalytical properties were determined for the electrodes having different oxide loading obtained at 450°C.

INTRODUCTION

Titanium electrodes coated by thin film of conductive d-group metal oxides, generally known as DSA® (Dimensionally Stable Anode), are widely employed in various areas of industrial electrochemistry as insoluble anodes: chlor-alkali industry[1], anodic oxidation of organic compounds [2,3], precious metal coating industry, decomposition of toxic organic wastes, and etc., due to their high corrosion resistance, good conductivity[4,5] and also showing lower overpotentials for the chlorine and oxygen evolution reactions which bring in advantageous to energy consumption reduction[2,4,6].

Various kinds of these type ruthenium oxide basis anodes were developed, such as: RuO_2[3,6-13], SnO_2-RuO_2-IrO_2[2], RuO_2-Co_3O_4[5], RuO_2-SnO_2[4,14], $Ru_xIr_{1-x}O_2$[15] and RuO_2-Ta_2O_5[16]. RuO_2 containing anodes have been reported to possess highest electrocatalytical activity towards anodic evolution reactions of oxygen and chlorine.

Several coating procedures are applied to prepare this kind of ceramic coatings; conventional thermal decomposition[3,5,6,17,18], sol-gel process[2,9,10,15], laser calcination[11], spray pyrolysis[12], sputtering[13]. To the best of our knowledge, there are few studies in the preparation of DSA®-type anodes by using Pechini method[4,7,8,14]. These investigations have revealed that Pechini method could be an alternative way of preparation providing better electrocatalytical and mechanical properties. The method involves similar coating steps with the conventional thermal decomposition procedure and differs from the precursor solution in which the metal ions distributed in a polyester network and also application of intermediate temperatures in order to improve the coating adhesion.

In the present study, titanium supported RuO_2 coatings were prepared by using Pechini method. The influence of decomposition temperature and the number of coating layers on the morphology, microstructure and electrochemical properties were investigated.

EXPERIMENTAL PROCEDURE

Substrate

Titanium plates (1 × 10 × 1 mm) were used as substrate, which were initially degreased in boiling HNO_3 at 90°C for 1 h then pickled in HF:HNO_3:H_2O solution with a ratio of 1:1:10 (v/v) for a few seconds and washed with distilled water. After that, they were sandblasted followed by an Ultrasonic cleaning in acetone for 15 min. Before dipping in the precursor solution, the plates were etched in boiling 3 M HCl in order to remove any insulating oxide film, washed with distilled water dried in the air and finally kept in isopropanol.

Precursor Solution

The coating solution was prepared dissolving citric acid (Merck) in ethylene glycol (Merck, extra pure) at 60°C. $RuCl_3.xH_2O$ (Heraeus) was slowly added into this mixed solvent and stirred. The mixture was diluted in isopropanol and acidifed by 5 ml of HCl (Merck). The Citric Acid:Ethylene Glycol:Ru molar ratio was 0.76:3.52:0.25.

Coating Procedure

Titanium substrates were coated by dip-withdrawal technique with a dipping rate of 20 mm/min. The material initially dried in air then treated at an intermediate temperature (~250°C) for 60 s to evaporate the solvent and improve the adherence of the coating. After that, it was treated at higher temperatures (400-450-500°C) for 10 min to remove the organic materials and to decompose the chloride into oxide form. Final coatings were kept in determined temperature for 1 h in order to complete the decomposition of the active metal into oxide form.

Physical Characterization

The physical characterization of the coatings was carried out by thin film X-ray diffraction analysis in the 2θ range from 20 to 80 at an increment of 0.5°. The morphology of the surface was examined by scanning electron microscope.

Electrochemical Measurement

The anode activity in OER was investigated by potentiodynamic (0.5 mV/s) polarization tests under N_2 flux in 1.0 M aqueous H_2SO_4 solution at 25°C. The anodes initially were pre-polarized for 1 h at constant current density (j = 20 mA/cm^2) to stabilize the surface by chronopotentiometry and immediately after that linearly sweeped in the potential range between 1.15 and 1.40 V (SCE). Prior to sweep, the anodes were kept at 1.15 V (SCE) for 120 s. Platinum foil (4 cm^2) was used as counter electrode, a saturated calomel electrode (SCE) was used as the reference electrode. Working area of the anode surface was 0.7 cm^2.

RESULTS

Physical Characterization of RuO_2 Films

In the physical investigation of the RuO_2 films coated on titanium, it is found that the phase ratios of Ti/RuO_2 changes with decomposition temperature.

Figure 1 shows XRD patterns of the RuO_2 coating after annealing at various temperatures. It is clearly seen in the figure that peaks having highest intensity belong to the titanium base metal whereas rutile-type RuO_2 peaks match much more peak with lower intensity.

RuO_2 peaks tend to become narrower with increasing temperature, which shows that particle size becomes larger. On the other hand, the RuO_2 peak intensity increased with further increase of the decomposition temperature, revealing that the RuO_2 gradually crystallized.

XRD patterns of the coatings also indicate that there is no $TiRuO_x$ phase established. Although there is a probability of formation of transition phases between the base metal and the coating, that couldn't be seen, since the XRD isn't taken from the cross-section but from the surface.

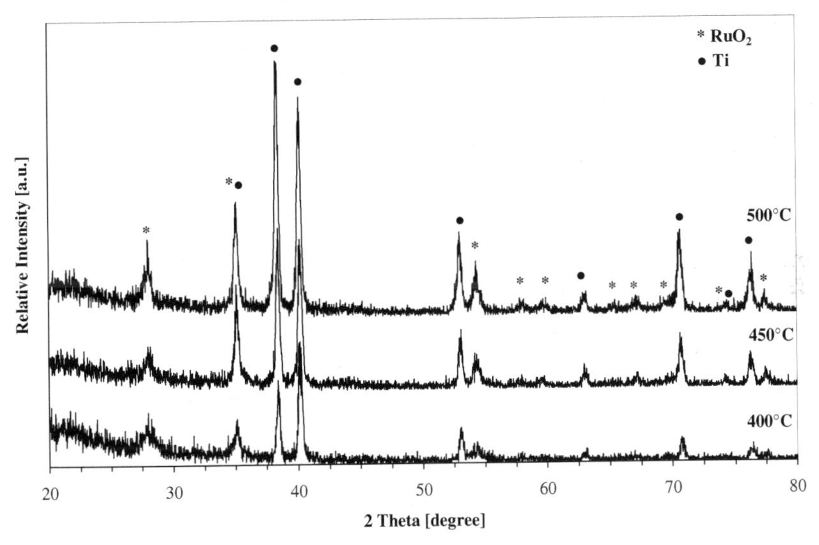

Figure 1. XRD patterns of 10 layer RuO_2 coating on titanium substrates calcined at various temperatures

In order to examine the oxide formation through the decomposition process of $RuCl_3.xH_2O$, morphological investigation was conducted by SEM (Fig.2). The coating is heterogeneous having typical "mud-cracked" morphology, containing agglomerates of RuO_2 crystals.

It can be observed that higher temperature favors the roughness of the electrodes by causing more flaws on the boundaries of flat-islands. These flaws having nano-sized RuO_2 fine particles (Fig. 2 b- 2 c) may be thought to possess better electrochemical properties due to higher effective surface area, although it was seen that there is no advantage of having nano-sized structures when used as oxygen evolving anode due to the fact that there was no identified slope changes of Tafel line. At higher decomposition temperatures the formed layers showing heterogeneous morphology consist of both mud-crack flats and their agglomerated form. This structure indicates that crystallization and agglomeration start when the spaces between the islands disappear and decomposition products begin to accumulate on the boundaries of islands.

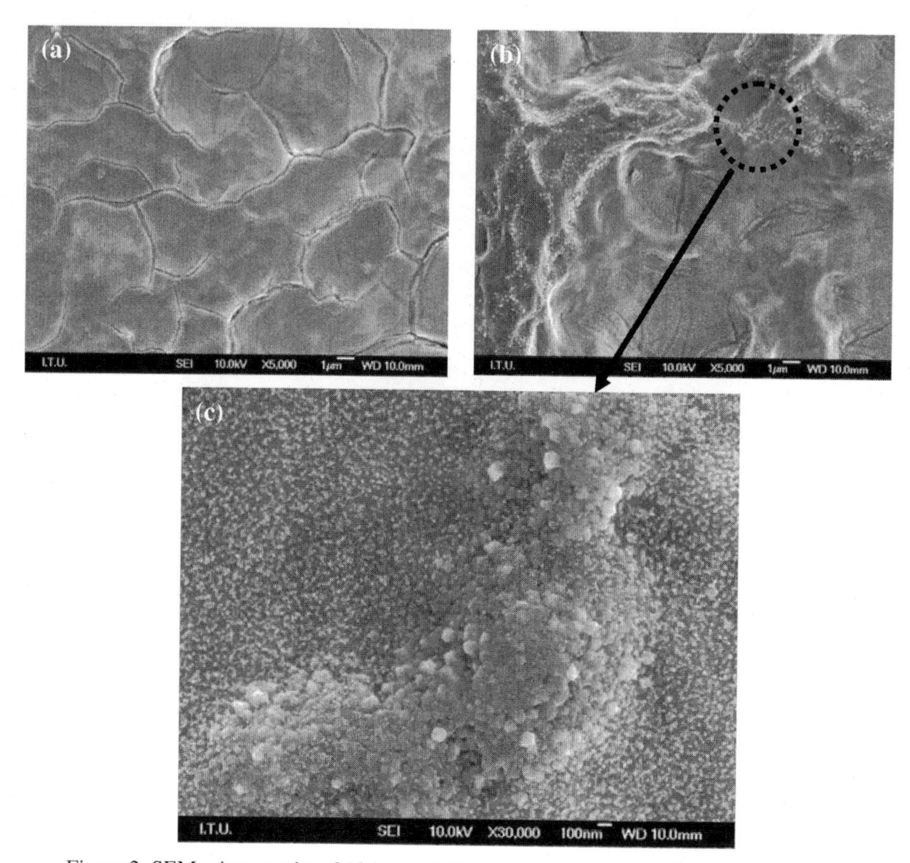

Figure 2. SEM micrographs of 10 layer RuO_2 coating on titanium substrates calcined at (a) 400°C (b)-(c) 450°C

In Figure 3, 1 and 3 layers of RuO_2 coatings were shown, which demonstrate the common surface morphology of "mud-crack" flats of islands. As the coating layer got thicker, the spaces (crevices) between the flat-islands became narrower and in most case spaces disappeared due to the multi-layered coatings covering those areas. It is also in good agreement with the micrograph of 10 layers of the metal oxide coating given in Figure 2 (b), showing almost no spaces between the islands.

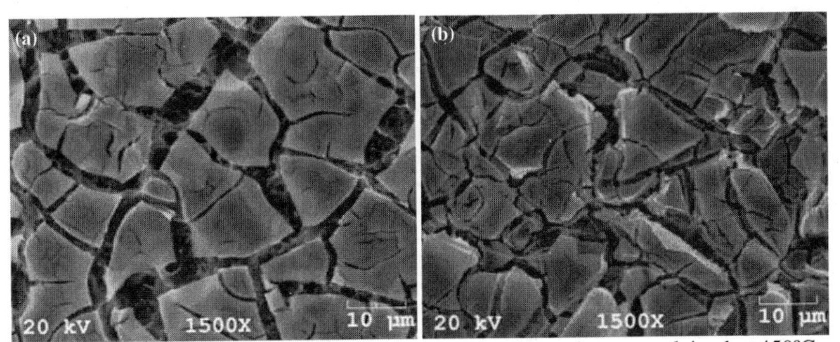

Figure 3. SEM micrographs of RuO_2 coating on titanium substrates calcined at 450°C (a) 1 Layer, (b) 3 Layers of RuO_2.

Electrochemical Measurement

The oxide coatings were conducted to 24-h-electrolysis under simulated industrial plating applications (20 mA/cm^2, 1.0 M H_2SO_4, 25°C). In Figure 4 the potential/time curves of the anodes with various coating layers prepared at 400°C was given; anode potential of 1 layer coated RuO_2 increases after 2.5 h whereas 10 layers of RuO_2 remains the same, showing that the coating mass is directly related to the anode potential and favors the stabilization in prolonged electrolysis time. As the numbers of coatings increase, the surface becomes more resistant to the experimental conditions.

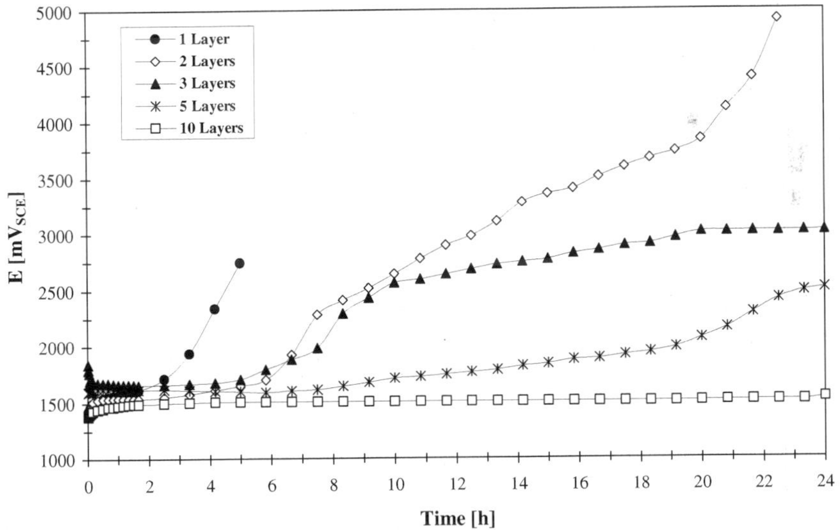

Figure 4. Potential change with time; anodes prepared at 400°C (1.0 M H_2SO_4, j =20 mA/cm^2, N_2-flux, 25°C)

No significant change in the anode potential is seen, in the case of 10 layers of RuO_2, also indicates that using these type anodes, higher coating mass is an inevitable requirement. Since the aim of industrial application is to reduce and keep the cell potential to a minimum[1], several numbers of coatings are beneficial.

Potentiodynamic polarization tests were applied to the anodes prepared at different temperatures and the polarization curves and Tafel plots are given in Figure 5.

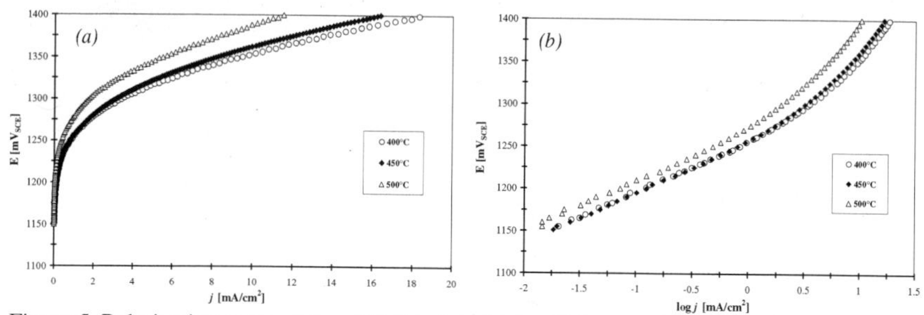

Figure 5. Polarization curves (a) and Tafel plots (b) for 10 layers of RuO_2 coatings for OER in 1.0 M H_2SO_4; (N_2-flux, ν= 0.5 mV/s, T = 25°C)

As it can be seen in the figure, there are no significant difference between the ones prepared at 400°C and 450°C. At lower current densities the Tafel slopes were found to be ~60 mV for all temperatures. But when considering a specific current, the anodes show lower potentials in the case of prepared at lower temperatures.

It has been reported that a Tafel slope of 81 mV could be observed for the RuO_2 anodes prepared by Pechini method at 400°C[14]. In an another study[10] where the anodes were prepared by the alkoxide sol-gel procedure at 450°C, the Tafel slopes for the OER were reported to be around 40 mV and the authors said that the anode activity in OER is definitely independent of coating mass. Active component of the coating (RuO_2) controls the reaction mechanism since all anodes having different coating masses have similar Tafel slopes. In other words, the anodic reaction takes place through the active site of the surface. So it is considered that coating mass may not affect the electrocatalytical activity in OER but it should be taken into account when prolonged and severe anodic polarization conditions utilized.

Tafel plots of anodes obtained at 400°C are shown in Figure 6. For a given current density, material coefficient (a) is lower for the RuO_2 electrodes having highest number of coating layers. Although electrocatalytical activity is scrutinized through b-values, it was seen that all electrodes with different coating masses show similar b-values. Therefore, it should be noted that coating mass does not affect the electrocatalytical activity of the electrodes but allows the material coefficient (a) become lower.

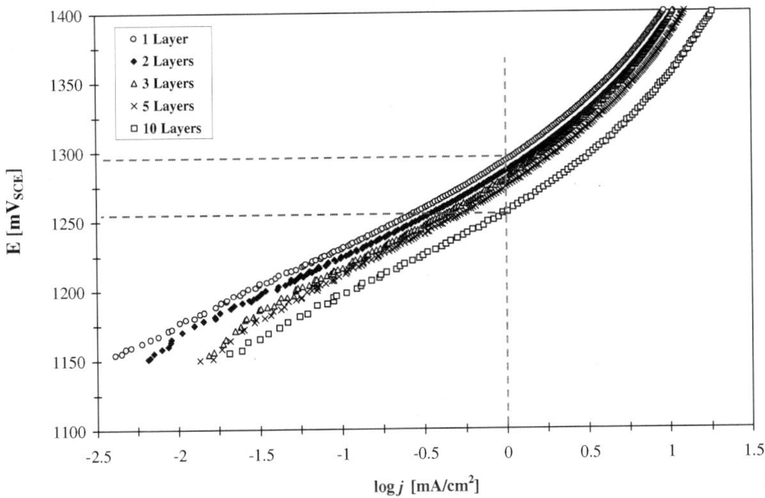

Figure 6. Polarization curves for RuO_2 coatings calcined at 400°C.
(N_2-flux, 1.0 M H_2SO_4, v= 0.5 mV/s, T = 25°C)

In Table 1 Tafel slopes and the anode potentials after 1 h-constant anodic polarization (j = 20 mA/cm^2) of the different numbers of coatings prepared at 450 and 500°C are given. There are two Tafel slopes for each anode, indicating a change in the reaction mechanism with the polarization (depending on the change of current density); the former is obtained at E < 1250 mV and the latter is obtained at E >1250 mV. Higher decomposition temperatures cause higher tafel slopes and lower activity for the OER. It is mentioned in Ref.[1] that RuO_2 electrodes, prepared at increasing temperatures exhibit an increasing Tafel slope for O_2 evolution.

Table 1. Tafel slopes and anode potentials after 1-h-anodic polarization at j= 20mA/cm^2

Number of Coatings	Tafel slope (b, mV/dec)	$E_{1\ hour}$ (mV)	Tafel slope (b, mV/dec)	$E_{1\ h}$ (mV)
	450°C		500°C	
1	56 79	1586	99 142	1957
2	55 87	1560	67 109	1674
3	59 97	1522	67 114	1690
5	61 101	1507	69 111	1635
10	60 107	1444	60 109	1583

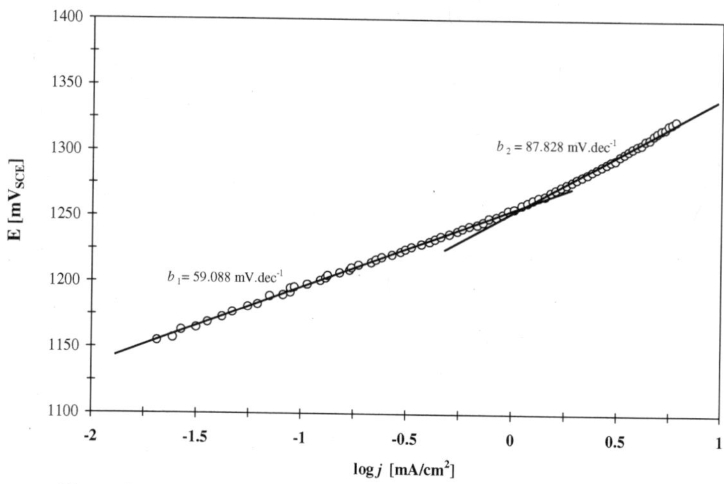

Figure 7. Tafel plot for 10 layers of RuO_2 coating calcined at 400°C. (N_2-flux, 1.0 M H_2SO_4, ν= 0.5 mV/s, T =25°C)

In Figure 8, Tafel line of 10 layers of RuO_2 coating was given. At lower current densities Tafel slope was found to be ~59 mV. When the polarization increases, a break in the Tafel line was observed above 1250 mV with a slope of ~88 mV which indicates a change in the reaction mechanism. Regarding this change in *b*-values it is observed that; at lower overpotentials, $2e^-$ transfers and at higher overpotentials $3e^-$ transfers are required for 1 mol of O_2.

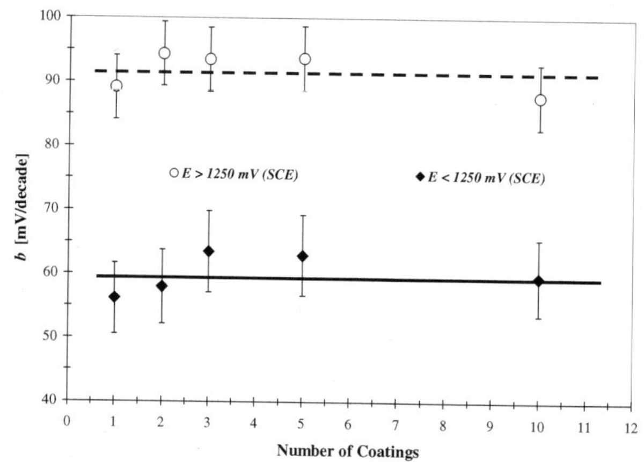

Figure 8. Tafel slopes for the different coating masses calcined at 400°C. (N_2-flux, 1.0 M H_2SO_4, ν= 0.5 mV/s, T =25°C)

Tafel slope of 63 mV at low overpotentials and 97 mV at high over potentials were found in literature which were related to possible surface changes for the former and an increase in the strength of the adsorption of the OH intermediate on the active site for the latter[17]. It is reported that as the particle size of RuO_2 on the film increases, Tafel slope increases to 60 mV[18]. It is more acceptable when considering the *b*-values obtained in this work.

Two different *b*-values were obtained for each electrode. At lower overpotentials anodes exhibit better electrocatalytical activity for OER. With increased polarization, $2e^-$ were transferred for the OER. Based on these results, RuO_2 anodes prepared by Pechini method are used in conventional plating applications at low current densities by means of possessing lower oxygen depolarization. At high current densities, these electrodes are thought to be preferred as active chloride electrode in chloride containing electrolyte where the OER depressed, although behaviors of these electrodes in chloride containing electrolyte haven't been investigated in this study yet.

CONCLUSIONS

Electrocatalytically active Ti/RuO_2 anodes are prepared by Pechini method. Higher decomposition temperature favors the crystallization of RuO_2 nano particles which also changes the surface morphology from mud-cracked to more agglomerated morphology. Prolonged electrolysis under constant current densities reveals that more stabile anodes in acidic electrolyte are the ones having higher coating masses and prepared at lower annealing temperatures. On the other hand, Tafel slopes of ~59 mV are obtained with the anodes prepared at both 400°C and 450°C with different oxide loading, showing that the anode activity in the OER are independent of coating mass. Decomposition temperature of 400°C and coating number of 10 layers are suitable for the calcination process. For a given range of current density, ($j = 0 - 5$ mA/cm^2) ΔE of anodes were found to be 66, 71 and 98 mV for 400°C, 450°C and 500°C, respectively which indicates that lower decomposition temperatures are more suitable for the improvement of the electrochemical properties of this kind of electrodes.

ACKNOWLEDGMENTS

The financial support of The Scientific and Technological Research Council of Turkey (Project No. 106M249) is gratefully acknowledged.

*: Corresponding author. Phone: +90 212 2853369; fax: +90 212 2853427.
E-mail address: kahveciog3@itu.edu.tr

REFERENCES

[1] S. Trasatti, Electrocatalysis: Understanding the Success of DSA®, *Electrochim. Acta*, **45**, 2377-2385 (2000).

[2] M.E. Makgae, C.C. Theron, W.J. Przybylowicz and A.M. Crouch, Preparation and Surface Characterization of Ti/SnO_2–RuO_2–IrO_2 Thin Films as Electrode Material for the Oxidation of Phenol, *Mat. Chem. Phys.*, **92**, 559-564 (2005).

[3] G.R.P. Malpass, R.S. Neves and A.J. Motheo, A comparative Study of Commercial and Laboratory-Made Ti/$Ru_{0.3}Ti_{0.7}O_2$ DSA® Electrodes: "In situ" and "Ex situ" Surface Characterization and Organic Oxidation Activity, *Electrochim. Acta*, **52**, 936-944 (2006).

[4] X. Wang, D. Tang and J. Zhou, Microstructure, Morphology and Electrochemical Property of $RuO_2$70$SnO_2$30 mol % and $RuO_2$30$SnO_2$70 mol% Coatings, *J. All.& Comp.*, **430**, 60-66 (2007).

[5] L.M. Da Silva, J.F.C. Boodts and L.A. De Faria, Oxygen Evolution at $RuO_2(x)$ + $Co_3O_4(1-x)$ Electrodes from Acid Solutions, *Electrochim. Acta*, **46**, 2001, 1369-1375 (2001).
[6] J. Aromaa, and O. Forsén, Evaluation of the Electrochemical Activity of a Ti–RuO_2–TiO_2 Permanent Anode, *Electrochim. Acta*, **51**, 6104-6110 (2006).
[7] A.J. Terezo and E.C. Pereira, Preparation and Characterization of Ti/RuO_2 Anodes Obtained by Sol-Gel and Conventional Routes, *Mat. Let.*, **53**, 339-345 (2002).
[8] L.A. Pocrifka, C. Gonçalves, P. Grossi, P.C. Colpa and E.C. Pereira, Development of RuO_2–TiO_2 (70–30) mol% for pH Measurements, *Sensors&Actuators B*, Vol **113**, 1012-1016 (2006).
[9] V.V. Panic, A. Dekanski, S.K. Milonjic, R.T. Atanasoski and B.Z. Nikolic, RuO_2–TiO_2 Coated Titanium Anodes Obtained by the Sol-Gel Preocedure and Their Electrochemical Behaviour in the Chlorine Evolution Reaction, *Coll.&Surf. A: Physicochem. Eng. Asp.*, **157**, 269-274 (1999).
[10] V.V. Panic, A.B. Dekanski, S.K. Milonjic, V.B. Miskovic-Stankovic and B.Z. Nikolic, Electrocatalytic Activity of Sol–Gel-Prepared RuO_2/Ti Anode in Chlorine and Oxygen Evolution Reactions, *Russ. J. Electrochem.*, **42** (10), 1055-1060 (2006).
[11] R.R.L. Pelegrino, L.C. Vicentin, A.R. De Andrade and R. Bertazzoli, Thirty Minutes Laser Calcination Method for the Preparation of DSA® Type Oxide Electrodes, *Electrochem. Comm.*, **4**, 139-142 (2002).
[12] J.L. Fernandez, M.R.G. De Chialvo and A.C. Chialvo, Ruthenium Dioxide Films on Titanium Wire Electrodes by Spray Pyrolysis: Preparation and Electrochemical Characterization, *J. App. Electrochem.*, **27**, 1323-1327 (1997)
[13] L.M. Doubova, A. De Battisti, S. Daolio, C. Pagura, S. Barison, R. Gerbasi, G. Battiston, P. Guerriero and S. Trasatti, Effect of Surface Structure on Behavior of RuO_2 Electrodes in Sulphuric Acid Aqueous Solution, *Russ. J. Electrochem*, **40**, 1115-1122 (2004).
[14] J.C. Forti, P. Olivi and A. R. De Andrade, Characterization of DSA®-Type Coatings with Nominal Composition $Ti/Ru_{0.3}Ti_{(0.7-x)}Sn_xO_2$ Prepared via a Polymeric Precursor, *Electrochim. Acta.*, **47**, 913-920 (2001).
[15] F.I. Mattos-Costa, P. De Lima-Neto, S.A.S. Machado and L.A. Avaca, Characterization of Surfaces Modified by Sol-Gel derived $Ru_xIr_{1-x}O_2$ Coatings for Oxygen Evolution in Acid Medium, *Electrochim. Acta*, **44**, 1515-1523 (1998).
[16] J. Ribeiro and A.R. De Andrade, Investigation of the Electrical Properties, Charging Process, and Passivation of RuO_2–Ta_2O_5 Oxide Films, *J. Electroanal. Chem.*, **592**, 153-162 (2006).
[17] D.T Cestarolli and A.R. De Andrade, Electrochemical and Morphological Properties of $Ti/Ru_{0.3}Pb_{(0.7-x)}Ti_xO_2$ – Coated Electrodes, *Electrochim. Acta*, **48**, 4137-4142 (2003).
[18] E. Guerrini, and S. Trasatti, Recent developments in understanding factors of electrocatalysis, *Russ. J. Electrochem.*, **42** (10), 1017-1025 (2006).

THE MYRIAD STRUCTURES OF LIQUID WATER: INTRODUCTION TO THE ESSENTIAL MATERIALS SCIENCE

Rustum Roy[1, 2] and Manju L. Rao[1]

[1]The Pennsylvania State University, University Park, PA 16802, USA
[2]Arizona State University, Tempe, AZ 85287, USA

ABSTRACT

The significance of water to life, including humans and human society and human technologies including energy, surpasses by orders of magnitude that of any other single material that MST members will ever work with. Yet, by and large, water has been left to chemists to study, and neglected by materials researchers. This paper focuses on the materials scientists' contribution to defining the structure, and the structure- property relations of water and ultra-dilute aquasols including all natural waters.

The "black holes" in the materials science as we have learned and taught it, include: truncation of the number of easily accessed intensive variables in our thermodynamics textbooks from P, T E and H, to only P and T. Hence the gross neglect of the remarkable influence of radiation (acoustic and EM) on matter, especially at those special and rare frequencies which cause resonance in the system at various levels.

We present a summary of the 2008 version of the leading physicists' and chemists' views of the structure of water. We then summarize our own recent materials science perspective on the structure of water and present the substantive body of experimental evidence largely, but not only, from our laboratory, of some truly extraordinary changes caused by these new vectors on the structure of water, including some radically new phenomena such as the burning of water and imprinting of water by highly specific frequencies, and the creation of the myriad new (albeit poorly defined) and possibly metastable structures of water.

INTRODUCTION

The Materials Science Approach to the Structure of Water

It should be unnecessary to introduce water to any scientist, but perhaps it is not amiss for those of us in ceramics, metallurgy and polymer science. None of us considers water as part of our territory! Yet water in its liquid and solid states are the 1st and 2nd most abundant mineral phases on our earth. While the structure of solid water—ice, has been well studied for decades by materials scientists-physicists, and even glassy water, a rare laboratory creation, has received a great deal of attention from them, liquid water has, almost unbelievably, been neglected. Of course, part of that is due to the enormous difficulty of dealing with the structure of aperiodic matter. One might have expected that materials scientists could devote extraordinary energy to overcome these challenges if one considered the significance of dealing with the single phase which outranks all other materials by orders of magnitude in its importance to human life. One part of the explanation for this omission is that "water" as a material was considered to be part of the "territory" of chemists, not of materials scientists. This paper, therefore, is a beginning to rectify that situation. Liquid water is the world's most important (ceramic) material by several orders of magnitude and only condensed matter, or materials, scientists can do justice to the study of this phase.

The fine detailed studies by chemists have indeed contributed to the detailed understanding of the composition and chemical reactions of, and in, water. And masterful analyses have been made of the "structure of water" and that is the source of today's problem. Michael Faraday, Joseph Priestley and Antoine Lavoisier in various iterations around 1831 gave us the H_2O composition correctly. The problem started with the fact that water was a very rare composition of matter that could be easily produced and easily put to use in all three states—solid, liquid and gaseous. It was but a small slip thence, to begin to equate water with H_2O. The structure of water equated then to the "Structure of H_2O." H_2O as a molecule is present only in the gas. But H_2O is also used to describe liquid water, which is a condensed phase and its structure can only be described in the appropriate metrics and language for condensed phases, very different from those for molecules.

There is no better illustration of the saying "a little learning is a dangerous thing!" 'Everyone' since the 19th century knew that water was "just water," hence, "everyone" believes that since water is JUST H_2O; i.e. always - just the same H_2O- hence the term "structured water" is obviously absurd. It is after all JUST H_2O. The confusion caused, worldwide, and society wide, for well over a century equated the fact of the composition being always the same—JUST H_2O, with the structure being always the same, and this has had profoundly negative consequences in Science and medicine as we will show. The confusion starts, but does not end, with the term "structure." What does this mean to the non-specialist? It can apply of course to the molecular level—the shape and linkages and interactive distances among atoms, often analyzed in a vapor. But we, here, are focusing now on condensed matter: liquid or solid, focusing largely on the liquid phase. To a materials scientist, the structure of a condensed phase—the periodic condensed phase, is the knowledge of the position of each atom in the 3D crystal. This can be known with great precision—such as 0.001Å—in most crystalline phases today. It can be determined by X-rays, neutron and electron diffraction, and confirmed by direct imaging by HRTEM, seeing literally atom by atom. However, when we enter the world of '*aperiodic*' matter; i.e. liquids and glasses (frozen liquids), such precision vanishes. But the meaning of the term "structure" to a materials scientist remains the same—the 3D positions of each particular atom or molecule, and secondarily the bonds between atoms or molecules holding them together, and recognizing the unavoidable Brownian motion effects on any structure in a liquid.

For determining the structure of liquids one has to resort, largely to spectroscopy of various kinds which tell one, statistically, of the bond lengths and distances of nearest neighbors, next nearest neighbors and further neighbors, etc. Moreover, the bonds between atoms which are associated inside a molecule and those holding molecules together can be very, very different. The bonds are nearly equal in completely ionic (like NaCl) or metallic (Fe) liquids to those in the same solids. Hence, bonding in a liquid can make a profound difference to structure. Those materials with equal bonds are called isodesmic, and unequal bonds, anisodesmic. And degree of anisodesmicity makes this profound difference. Liquid structures of NaCl and Fe for, example near the melting point look similar to the crystal with bigger amplitude in the simple harmonic motion of the ions or atoms. Anisodesmicity however can cause profound differences within two crystalline phases with exactly the same structure. We illustrate this with an easily and most clearly understood example, because it is confirmed by personal experience, repeated over and over by everyone. It has been very effective in changing—wherever known—the widely held belief that water must be the same, "just water," if it is really pure compositionally. The element carbon is well known worldwide in two forms—'graphite,' the softest (experienced whenever we write with a pencil) and diamond, the hardest material known. These are both composed 100% of carbon; just carbon, "JUST C." Obviously, this useful illustration proves to everyone that just having the same composition is no indication whatsoever of a unique set of properties for any condensed matter material. Thus, "just water," is just inadequate!

The diamond-graphite example has two other important implications. Firstly, if not composition, to what must we ascribe what is evidently a massive difference in properties? The obvious answer is what we accept as the "1st law of Materials Science." Properties reflect, and are determined by, structure. Hence, we know that it is structure that is paramount in determining the properties we encounter. The third useful illustration in this diamond-graphite example is the importance of bonding in determining and reflecting properties. In both graphite and diamond we have very strong covalent bonds between two carbon atoms. In graphite, they are sp^2 bonds lying in the plane of a sheet, in diamond, they are sp^3 covalent bonds forming tetrahedra in a continuous 3D network. The extreme softness of graphite arises from the anisodesmicity—unequalness of bonds—involved in the structure. While the bonds in the plane of the graphite sheets are indeed even stronger than in diamonds, the interatomic distances (inversely related to hardness) are 1.54 Å in diamond and only 1.43 Å in graphite within the sheet, but 3.35 Å between the sheets. The bonds between the sheets—van der Waals bonds—are extremely weak (bond length of 3.35 Å). The thoroughly studied carbon example provides the blueprint to argue the case that in all condensed phases, it is the weakest bonds which provide the most interesting properties. Hence, in all anisodesmic phases like water, it will be the weakest bonds that must be studied in detail. Being more difficult, this also makes them less studied.

a) The "black hole" in ceramic sciences. The near total neglect of liquid water:

Obviously, water is a ceramic. It is by orders of magnitude the most abundant component on the earth. It is at least also the most important phase on the planet, and uniquely significant scientifically as the *sine qua non* for life. Yet it escapes the attention of all younger scientists. This neglect has led to the misleading use of the term "structure of water" by those trained in chemistry and the consequent "disaster" in major sections of science, especially those connected to living systems, especially modern medicine. The chemists' use of the term "structure of water" referred exclusively to the structure of the molecules in the liquid water. None of it dealt with the "structure of water" as material scientists use that term: the arrangement of the atoms in 3D space. Yet the elegant computer drawn images of molecules have confused an entire generation of scientists from theoretical physicists to biologists and most consequentially, medical researchers far removed from these precise distinctions, into substituting in their minds, these beautiful images for the real structure of liquid water. Prof. Martin Chaplin

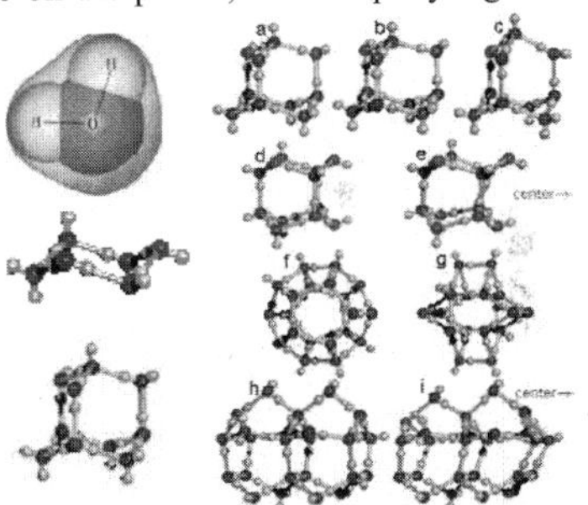

Figure 1: The enormous variety of structures of the molecules in which almost certainly the chemical entity H_2O can exist. The well known H_2O monomer with its precisely defined tetrahedral angle is shown on the top left and below it a series of dimers, trimers, tetramers, etc.(1)

is without doubt, the world authority on the molecular structures of water and Figure 1 shows an array of monomers,—the classic rabbit-eared H_2O—and several calculated structures of oligomers of H_2O (taken from his work).[1]

In 2005, the senior author and his colleagues presented the first thorough analysis of the "crystal" structure description of water to which the reader is referred to for details.[2] Of course, limited by the fact that it is a liquid and relatively fluid, it is meaningless at this stage to attempt anything other than the concept of the structure of water as a nano-heterogeneous assemblage—determined by statistical mechanics (at equilibrium) of a mixture of oligomers. No one has any data on which combinations are present at any particular combination of all the relevant intensive variables: P, T, E, H; nor on the concentration of each, and certainly not on their distribution in space, or their stability or metastability. Our cartoon makes one much more general and significant point. In all such covalent liquids, as we have hinted above, the structures are strongly anisodesmic. The covalent bonds of O-H inside the oligomers are much, much stronger than the van der Waals bonds (black-black) holding different oligomers together as shown in Figure 2. Unfortunately, the figure cannot easily present the scaled spatial relations among the actual molecules, nor the probable clusters which are present; because no such data exist. The forces between the clusters outlined in black must be very much weaker than the intracluster forces, although the bond terminations are not drawn thus.

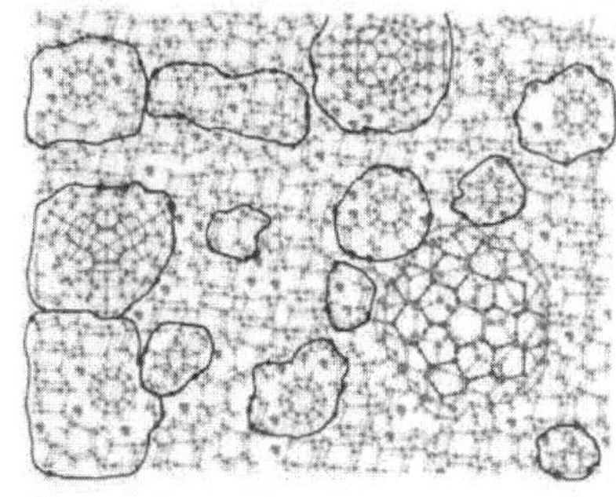

Figure 2: Cartoon of schematic presentation of the kind of space-filling mixture of molecular units which must exist in some proportion of smaller 2-4 molecule clusters (Fig.1) and other larger molecules up to the calculated 280 molecule units., to emphasize the key element of nano-heterogeneity of structure within water.

b) Some neglected branches of Science: The "black holes" in thermodynamics taught in Materials Science and physics

Although thermodynamics is routinely taught in chemistry and physics, it is a core of the learning which is specific to materials science—and which is increasingly diminished in favor of the fashion of the decade: superconductors, ceramic engines, room pressure diamonds and "nano-X." Examine if one will, the textbooks from which one learned thermodynamics. One of us (RR) taught graduate students about the Phase Rule (the most relevant part of thermodynamics for Materials Scientists) from the book by Ricci.[3] This and all other such books treated the stability of any phase judged by the free energy function; G. as dependent on two (and only two) intensive variables, pressure (P) and temperature (T). Of course, they are by far the most common, important and invaluable. One of us (RR) also spent 30 years of his career determining some of the most significant phase diagrams in ceramic science such as Al_2O_3-SiO_2, Al_2O_3-SiO_2-H_2O and BaO-TiO_2. And the only variables other than the chemical composition (an extensive variable) were always pressure and temperature, P, T.

But surely it was simply a truncation of thermodynamics to state only that:

$$G = H - TS \quad \text{and} \quad \frac{\partial G}{\partial T} = -S \qquad \frac{\partial G}{\partial P} = V$$

and we rarely even went on to the second derivatives, specific heat C_p, compressibility and thermal expansion.

$$\frac{\partial^2 G}{\partial T^2} = \frac{-C_p}{T} \qquad \frac{\partial^2 G}{\partial P^2} = -V\alpha \qquad \frac{\partial^2 G}{\partial P.\partial T} = \beta V$$

And when discussing phase transitions we used Ehrenfest's simple criteria[4] for a beginning. A first order transition is accompanied by a step function in the 1st derivative function of G; i.e. in volume or entropy. In a second order transition there are only changes of slope in those parameters and a step function in the second derivative of G. And so on. But now consider: are not electric fields (E) and magnetic fields (H), also intensive variables? Surely they are: and they can change properties. Hence, according to the first law of Materials Science they must change structure also. Materials scientists, if not most chemists and physicists, deal with ferro-magnetic and ferro-electric materials. At least in these cases, E & H play a profound role in major changes in properties. And further they interact directly with the other intensive variables. Temperatures change the magnetic (or electrical) properties drastically. It is clear that materials scientists have allowed a major error, a truncation of its most fundamental science to go unnoticed. At least four intensive variables are closely involved in changing the properties and structure of all matter in our everyday materials research. Of course, it is true that the changes are de minimis in the vast majority of ceramic and metals studied. But not in all. And that brings us back to water. We start with two examples of startling changes in solids and in water caused by E and H.

Crystalline, high temperature solid materials are dramatically affected by E & H fields:

During the period 1984-2007 our laboratory at Penn State has published over 200 papers and a dozen patents on the incredible structural changes which occur in any phase that contains unpaired electron spins. The effects include changes in reaction kinetics of 10^2-10^3; changing crystalline-"glassy" matter in the solid state, unpredicted microsctructure, all these effects are achieved at specific frequencies and typically in polarized fields.[4]

Dissociation of liquid water into H & O: Yes "water burns!"

This serendipitously discovered observation by John Kanzius, achieved by a polarized RF beam at≈ 13.56 MH z makes the importance of E & H fields effects on water immediately undeniable. But it goes much further. Since this requires the almost unimaginable: That a 5.2 eV O-H bond can be ruptured by ≈10^{-6} eV photon requires a new approach to physics which we attributed to the specific condition of resonance of the bonds involved at the frequencies which are effective. This also suggests that much early literature on the effects of weak E and H fields on water structure and properties cannot be discounted.

This conventional wisdom accepted by ceramic scientists worldwide through the 1950's that, "We cannot have different structures of the same pure liquid" was challenged first by Roy et al. in 1969, with relevance to the structure of liquids in general. They showed conclusively in their experimentally determined P-T phase diagrams that very common liquids can have stable P-T regions for different, liquid structures of fixed composition (specifically even monatomic, covalently bonded systems such as S, Se, Te, and even a partially metallic Bi).[5] Most of these have highly anisodesmic bonds (like water) best known in the chains of sulfur atoms. By a coincidence a month after the first presentation in April 2004, of the structure of water paper by Roy et al.,[6] a paper by Kawamoto et al.[7] appeared in which they presented an exact analogue to the P-T diagrams of S, Te, Bi diagrams, the P-T phase diagram for water showing at least one high pressure stable liquid region [Figure 3]. Thus, this experimental thermodynamic evidence confirms the plausibility argument adduced by us of the existence of at least some different stable structures in liquid water.

Further comments on metastable (in addition to stable) phases:

We must also record the fact that in all strongly bonded phases the probability of metastable phases is high. In solid SiO_2 the closest analogue (structurally) to water,[8] we have nearly an infinite number of tridymites, and cristobalites[9] and a literally infinite number of structure of glasses of SiO_2 composition.[2,10] In the case of carbon, when we include the anisodesmicity, we already know of the infinite variations of glassy carbons and recently of carbon nanotubes—all of exactly the same composition—C. Hence in water the number of structural possibilities are obviously similar.

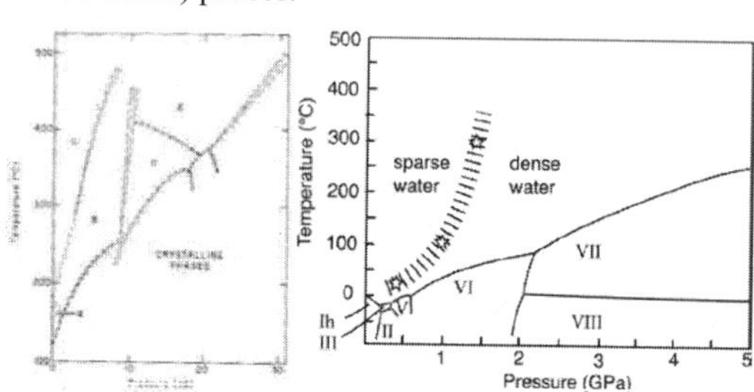

Figure 3(a): The Pressure-Temperature phase diagrams of liquid sulphur (6) [left]. **(b)** On the right is the Pressure- temperature phase diagram of water (a) showing the phase transition between low-pressure "sparse water", and high-pressure water "dense water" in the liquid stable water region, exactly analogous to the experimental data on several liquid structures such as in liquid

The entire matter of the existence of myriad - structures of water is essentially settled. The question that remains is, which can they be retained metastably and for how long and under what conditions of P, T, E & H. In materials science we retain such metastable phases by rapid "quenching" of the variable. In liquids the problem is that they most will usually freeze to crystalline matter. Albeit water, of course, has long been known to form glass just like its ceramic twin SiO_2. But how faithfully its structure will be retained in unknown. One set of metastable states of water have received an inordinate amount of attention in the chemical literature. These are the low temperature (<<0°C) glasses and inference from them can be drawn from them for liquid water. Angell[11] has provided a very thorough recent review of the subject. Can magnetic fields or electric field-quenching have similar effects? These are all huge areas for study on ordinary liquid water.

Current state of thermodynamic knowledge on the structure of water

The phase behavior of solid water, ice is now widely accept and well understood. The work of Tammann and Bridgman[12] and many others indicate that solid ice under different conditions of temperature and pressure forms different structures, and their work was intensively followed by Whalley and colleagues.[13] The present state of the solid water phase diagram is now well understood. Although various attempts have been made by scientists to explain the multiple anomalous properties of liquid water until the 1950s when physicochemical studies of water and its interactions with solutes gathered momentum and several molecular models were proposed. This gained scientific respectability in the 1960's when Frank[14] proposed the existence of long-lived structures in liquid water and aqueous solutions having (ion-solvent interactions) with the formation of 'flickering clusters' wherein the formation of one hydrogen bond is a co-operative phenomenon making and breaking many other hydrogen bonds.

Many others have contributed to the concept of various molecular - size "structures" present in varying concentrations in various waters under different P-T conditions.[15-18] The uniqueness of water is that it exhibits many such discontinuous changes in atleast 65 properties as identified by Chaplin[1] and Stanley[18] in its density, material, physical and thermodynamic properties, many of them in the 0-100°C liquid stable region on which we focus here. Among their "65 property anomalies," the most interesting ones are those that occur in the most significant thermodynamic functions, i.e. the first and second derivatives of the Gibbs function. These are illustrated in Fig. 4. *Compressibility*: In typical liquids, compressibility decreases as we lower the temperature, while in water the average isothermal compressibility is twice as large compared to a typical liquid and undergoes the most unusal non-monotonic change. The second significant function is the specific heat which is similar and twice as large as a normal liquid. The discrepancy gets larger as the temperature is lowered and changes sign at 35°C. The third 2nd order thermodynamic function is the coefficient of thermal expansion: Figure 4 shows the data on these specific thermodynamic functions of (a) *specific heat Cp,* (b) *isobaric cubic expansion* and (c) *isothermal compressibility* of liquid water as a function of temperature *T* taken from the review by Kumar, Franzese and Stanley.[18] The behavior of a normal liquid is shown as the dashed curve. It is only possible to explain the change in thermal expansion from a low positive to a high negative as the evidence for a substantial structural transition.

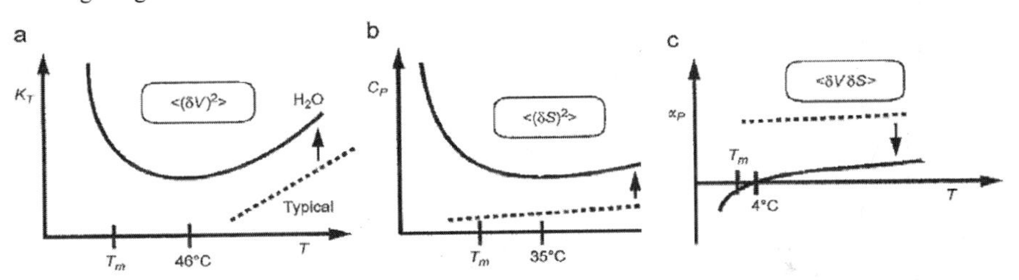

Figure 4: Schematic dependence on temperature of (a) isothermal compressibility K_T, (b) the constant- pressure specific heat, C_P, and (c) the coefficient of thermal expansion α_P. The behavior of typical liquid id indicated by dashed line, which, very roughly, is an extrapolation of the high-temperature behavior of liquid water. The anomalies displayed by liquid water are apparent about the melting point T_m, but are more striking below it.

Structure of water: Learning from water's nearest cousin, SiO_2

Materials Scientists have written tens of thousands of papers on SiO_2. Its many crystalline structures, perhaps 6 major, stable and hundreds of metastable ones are not relevant here. But SiO_2 –glass research is certainly the most relevant to water research. This enormous body of work is rarely mentioned in the chemical literature. We only select a few relevant learnings from the silicate-glass structure research: In 1960, Roy[19] first proposed the thermodynamic consideration which favored the tendency to phase separate in all systems with non-ideal Raoult's law, liquidus line. This fundamental concept was used to interpret and predict what compositions were likely to form the very valuable glass ceramics. Extremely small regions of nano-heterogeneity formed in the liquid made possible minute crystal formation without the usual nucleation and growth of crystals. This nano-heterogeneity of many glasses made from anisodesmic liquids challenged the ruling paradigm of the Zachariasen glass model.[20] In 1971, Roy[21] first proposed that this nano-heterogeneous model dominated the structure of many useful glasses and also other important covalently bonded liquids including, specifically water.

In 1986, when Porai-Koshits and Mazurin[22] demonstrated by their TEM data that the nano-heterogeneity model was not challengeable. With direct TEM images, they showed that many glasses consist of 2 or 4 separated phases. The concept of nano-heterogeneity was clearly established and has never been challenged; it has been ignored! At any given time liquid water exists as a statistical – mechanical thermodynamic equilibrium of multiple oligomeric units arranged in space. Bockris and Reddy[23] also provide cartoons as in Roy's 1971 paper requiring different size nano-units of structure.

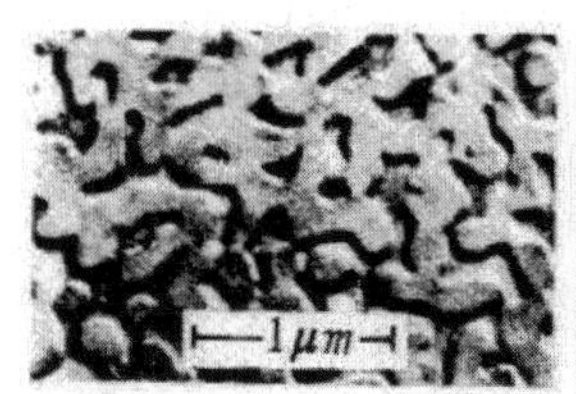

Figure 6: TEM images of binary, ternary and some quenched glasses clearly show actual phase separation, in sharp contrast to Zachariasen's random network theory. Structural (-compositional) fluctuations exist in most glasses and many liquids (22).

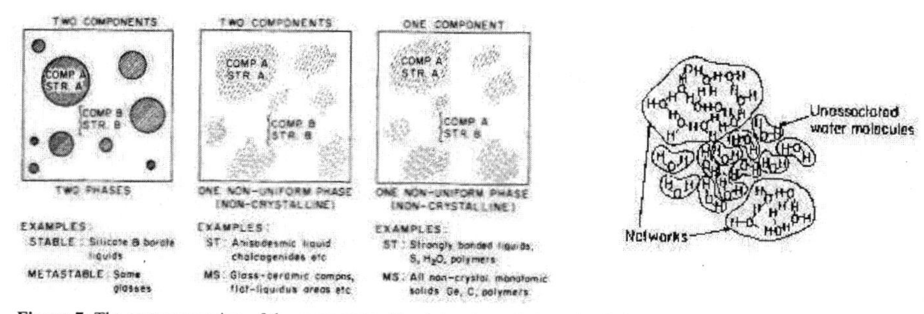

Figure 7: The cartoon version of the more generalized structure of glass clearly indicating its heterogeneous (with respect to structure or structure and/or composition (21). Note that water is mentioned in the third column. A similar representation of the water structure by Bockris and Reddy (23) is shown on the extreme right.

Chaplin's website shows numerous, splendid water clusters[1], but no 3D picture of the structures of the assembled liquid. Roy et al.[2] have proposed the only such crude generic nano-heterogeneous cartoon model for liquid water which at a given time exists as Chaplin's $(H_2O)_x$ trimers, oligomers and polymers where x varies from 2 - 250.

Experimental data on the Structure of water and aquasols:

A sol or colloid is a two phase system consisting of finely divided solid matter (< 100-1000 nm) permanently dispersed in a liquid. The finely divided phase in a stable colloid consists of (either positively or negatively) charged particles which prevent them from clustering together and / or precipitating out. Einstein, in his 1905 paper on Brownian motion, a typically colloidal phenomenon, and one of his three earthshaking papers of that year commented on the fact that 'colloids' are 'atoms' *structurally different from the parent liquid.* The colloidal state provides an excellent bridge to demonstrate the 'structuring effect' in water- the liquid phase. What is the nature and the influence of the charged solid-phase on the structure of the surrounding liquid phase, layer by layer, and the role of the structured cluster on it first, second and subsequent nearest neighbors? We have looked into the detailed analysis of silver aquasols in an attempt to answer some of these questions.

The colloidal state also provides an excellent bridge to demonstrate the biological effects on ultradiluted water samples. Metallic silver has been known for millennia and used for its exceptional antibacterial properties. Silver aquasols at 1-10 atom ppm concentrations are powerful broad spectrum antibiotic. The synergistic and additive affects of colloidal silver with various well established antibiotics has been well documented[24] For the study, we obtained our samples from several different companies that produce colloidal metal particles suspended in minuscule quantities (~ few ppm) in water.

We have analyzed the samples prepared by various technologies of 10 ppm, 30 ppm, 200 ppm and 400ppm of such silver aquasols. Further we have investigated particle size, morphology and the nature of the colloidal particles in waters. We have for the first time analyzed in detail the solid and the liquid phases using standard materials science analysis tools. The solid phase is analyzed using DTA, TGA, XRD. SEM and TEM, while the liquid phase was analyzed using FT IR, UV and Raman spectroscopy. We observe that there are atleast three different phases in the system: Ag-O: Ag, Ag_2O and Ag_4O_4 which are stable in air. In an aqueous environment between 0 and 100°C, there is evidence that various combinations of metal oxides, and possibly 'oxy-hydroxides' exist.[24]

Figure 8(a), shows the UV- VIS absorption spectra of a series of silver aquasols varying in Ag concentration, ranging from 10- 200 ppm. The UV absorption spectra was carried out in a double beam monochromatic UV- VIS spectrophotometer using silica glass cuvettes. While Ag Plasmon radiation in the UV absorption spectra occurs in the 400-420 nm regime, our analysis indicate major structural absorption effects due to the changes in the structure of water. Notable changes are also seen in the Raman spectra as indicated in 8(b). The presence of charged nanoparticles in water alters the bulk structure of water, by the influence of change (positive or negative) on the first nearest neighbors to the charges species, which in turn affect the second nearest neighbor and so on.

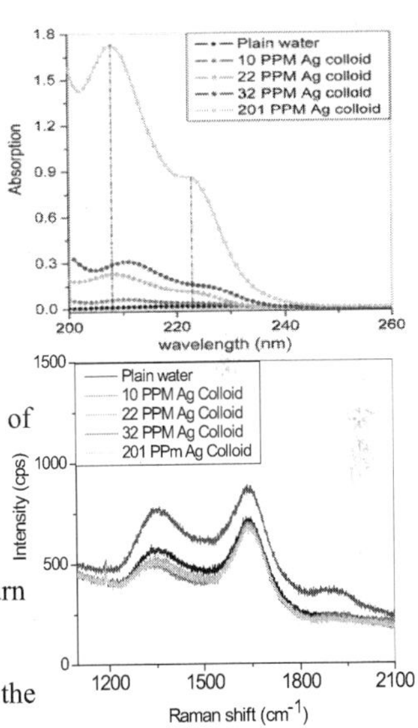

Figure 8: UV spectra of colloidal silver samples with varying silver concentration. Note their implication on the structure of water at lower wavelengths (200-240 nm). (b) Raman spectra of silver aquasols.

Effect of radiation on the structure of water:

A considerable body of work now demonstrates the effects of magnetic fields on aqueous solutions.[25] The influence of modest d.c. magnetic fields on the nucleation and growth of CaCO3 (phases, sizes, morphology) in dilute aqueous solutions have been thoroughly studied and demonstrated by Higashitani et al. and Pach et al.[26,27] The former demonstrates a strong memory effect in the constituent solutions exposed to the H-field. Tiller et al. have shown the remarkable effect of a static magnetic field on the pH of water in a conditioned space.[2]

We do not treat here, in any detail the rich patent literature on the use of RF and RF plasma transformation of water into new structures: only because the compositions of the phases are not carefully specified[28] The most dramatic effects of EM radiation came in a serendipitous discovery of Mr. J. Kanzius when he devised a circularly polarized radio frequency generator that was accidently found to be able to 'burn' salt water. Our initial studies using his device indicate that very low energy radio-frequency photons (~ millionth of an eV) are capable of and do break the O-H bond of water. A typical experiment involved the irradiation of NaCl solution in pyrex/quartz/Teflon test-tubes by the polarized RF fields. On ignition the water combusts with an obvious flame as shown in Fig 9. Various compositions and dilutions of NaCl ranging from 0.03 to 30 percent NaCl in water were investigated resulting in combustion of the gases, O, O_2 and H_2 liberated by the decomposition of the O-H bonds in H_2O.[30]

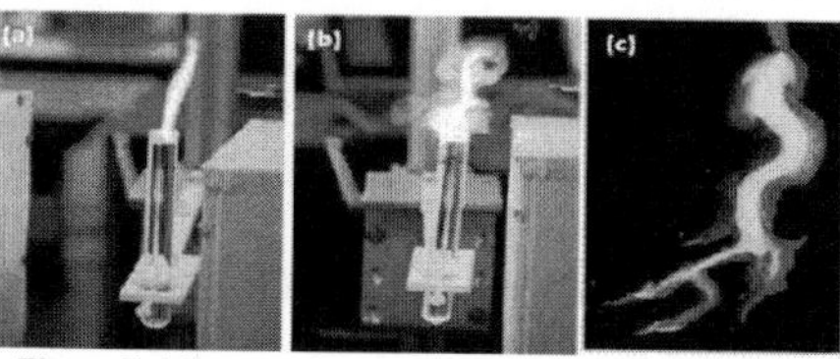

Figure 9: Mixtures of various concentrations of NaCl and water combusting (when ignited) in the presence of a 13.56 MHz RF radiation (a) 0.3% NaCl (b) 30% NaCl (c) droplet of natural sea water self-combusting in the RF fields.

An extremely stimulating result that sprang up form one of our experiments showed that spontaneous dissociation of water in the RF field leading to self – ignition as a drop of natural sea water, as it falls freely from tip of a micropipette. Figure 9c shows the spectacular image captured during the self-ignition of a drop of water in presence of the circularly polarized RF field. Structural analysis using Raman spectroscopy indicate that there is a definitive change in the structure of the pure NaCl-water solution after RF irradiation. Furthermore, it is important to note that the Raman spectral analysis of the saline solutions before and after the combustion confirms that there are substantial structural changes. These are not discussed further in this context, but show that such changes in the structure of the liquid phase are caused by radiation effects. The parallel influence of polarized 2.45GHz microwave fields on water using Sedlmayr's process[31] shows dramatic effects. The Raman spectra shown in Fig. 10 shows the changes in the intensity of the main OH stretch band of H_2O by an order of magnitude. It is interesting to note that such effects are metastable and relax over time, ≈7 hours in this particular example shown.

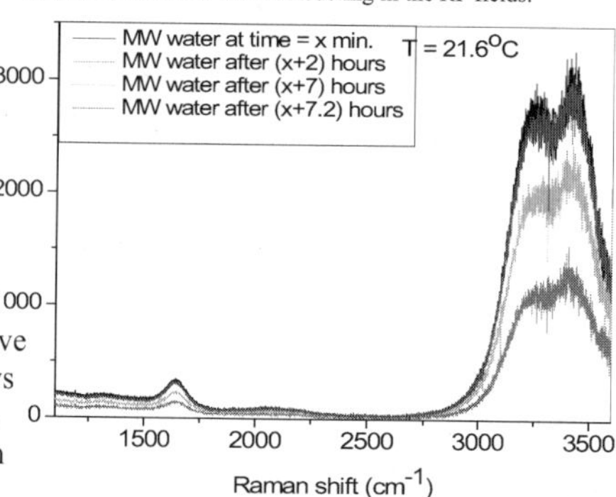

Figure 10: Raman spectra of microwave water measured at 21.6°C as a function of time. Note the time of relaxation for the structured-water is ~ 7.2 hours.

Although 2.45GHz and 13.56 MHz are typical frequencies in the microwave and radio frequency regions, it is interesting to note that unlike conventional fields, the radiation used in all these experiments is 'vectored.'

CONCLUSIONS

Drawing especially on the long and extensive work of Chaplin on the various molecular structures or clusters, and on the large amount of physical data on collected and collated by Stanley, to establish the large number of transitions in liquid water in the stable range, and largely on our own data on the spectroscopic data providing evidence for – structural changes produced by EM fields we believe that the following conclusions can be safely drawn:

1. Liquid water exists in dozens of different structures.
2. At one atmosphere pressure, the well known, stable, anomalous changes in properties provide the fingerprints of transitions between these thermodynamically stable phases. The P-T dependencies of such anomalies can provide useful thermodynamic parameters on such phase changes between specific structures.
3. In addition to our work, there is overwhelming empirical evidence especially in the patent literature that appropriate electrical and magnetic fields can cause structural changes in water, as they have been conclusively shown to do in high temperature solid matter.
4. Hence we believe that the materials science community has a great opportunity both to expand the science of the understanding and display of the structures of water, and to create and stabilize metastable structures of water which could no doubt open up major opportunities in the materials industries and in the world of human healing.

REFERENCES

[1]M. Chaplin, http://www.lsbu.ac.uk/water/intro.html

[2]R. Roy, W.A. Tiller, I. Bell, and M.R. Hoover, The Structure of Liquid Water; Novel Insights from Materials Research; Potential Relevance to Homeopathy, *Mater. Res. Innov.*, **9**, 577-608 (2005).

[3]J. E. Ricci, The Phase Rule and Heterogeneous Equilibria. Van Nostrand, New York, chap VIII, p 169 ff (1951).

[4]http://www.mri.psu.edu/centers/mpec/

[5]G.C. Vezzoli, F. Dachille, and R. Roy, Sulfur Melting and Polymorphism under Pressure: Outlines of Fields for 12 Crystalline Phases, *Science,* 166, 218-221 (1969).

[6]R. Roy, A Contemporary Materials Science View of the Structure of Water. FOH Symposium Science of Whole Person Healing, Sept 2004.

[7]T. Kawamoto, S. Ochiai, and H Kagi, Changes in the Structure of Water Deduced from the Pressure Dependence of the Raman OH Frequency, *J. Chem. Phys.,* 120, 5867-5870 (2004).

[8]J. D. Bernal and R. H. Fowler, A Theory of Water and Ionic Solution, with Particular Reference to Hydrogen and Hydroxyl Ions, *J. Chem. Phys.,* **1**, 515-548 (1933).

[9]Roy, D. M. and R. Roy, The Carnegieite-Nepheline and Cristobalite Tridymite Transitions, *Indian Mineral.,* 10:16-22 (1969).

[10]H.M. Cohen and R. Roy, Effects of High Pressure on Glass. The Physics and Chemistry of High Pressure, *Soc. for Chem. Ind., London,* (1962) pp.131-139

[11]C. A. Angell, Insights into phases of liquid water from study of its unusual glass-forming properties, *Science*, **319**, 582-587 (2008).

[12]P. W. Bridgman, Water in the Liquid and Five Solid Forms, Under Pressure, *Proc. Am. Acad. Arts. Sci.* **47**, 441-558 (1912).

[13]E. Whalley, D.W. Davidson, and J.B.R. Heath, Dielectric Properties of Ice VII. Ice VIII: A New Phase of Ice, *J. Chem. Phys.*, **45**, 3976–3982 (1966).
[14]H.S. Frank and W.Y. Wen, Ion-solvent Interaction. Structural Aspects of Ion-solvent Interaction in Aqueous Solutions: A Suggested Picture of Water Structure, *Disc. Farady Soc.* **24**, 133-140 (1957).
[15]P. G. Debenedetti, *Metastable Liquids* (Princeton University Press, Princeton, 1997).
[16]C. A. Angell, Formation of glasses for liquids and biopolymers, *Science* **267**, 1924 (1995).
[17]P. H. Poole, F. Sciortino, U. Essmann, H. E. Stanley, Phase behavior of metastable water," *Nature* **360,** 324-328(1992).
[18]P. Kumar, G. Franzese and H. E. Stanley, Dynamics and thermodynamics of water, *J. Phys.: Condens. Matter* **20** (2008) 244114-26.
[19]R. Roy, Metastable Liquid Immiscibility and Subsolidus Nucleation, *J. Am Ceram Soc* 43, 670-671(1960)
[20]W. H. Zachariasen WH, The atomic arrangement of glass, *J Am Chem Soc*, **54**, 3841-3851 (1932).
[21]R. Roy, Alternative to the random network structure for glass: non-uniformity as a general condition. *In: Hench L and Freiman SW (eds) Advances in nucleation and crystallization in glass. American Ceramic Society*, pp 57-60. (1971).
[22]Porai-koshits EA, V. I. Averjanov, Primary and secondary phase separation of sodium silicate glasses *J Non-cryst Solids* **1**, 29-38(1968); O V. Mazurin, E. A Porai-Koshits (1984). Phase separation in glass. North Holland, Amsterdam.
[23]J. O'Bockris, M. Reddy AKN Modern Electrochemistry, Vol 1, 2nd edn. Plenum Press, New York (1998).
[24]R. Roy, M. R. Hoover, A.S. Bhalla, T. Slawecki, S. Dey, W. Cao, J. Li, S. Bhaskar, Ultradilte Ag-aquasols with extraordinary bactericidal properties: the role of the system Ag-O-H_2O, *Mater. Res. Innov.*, **11**, 3-18 (2007). Also M. L. Rao, R. Roy and I. Bell, Characterization of the structure of ultradilute sols with remarkable biological properties, *Materials Letters*, **62,** 1487-1490, 2008.
[25]J.A. Prins (1929) Zeit, physik 56, 617- (199). Trans. Far. Soc. 33, 279- (1937)
[26]S. Duncan (1995). MS thesis in Mat. Sci. The Pennsylvania State University, University Park, PA.
[27]K. Higashitani, A. Kage, S. Katamura, K. Imai, and S. Hatade, *J. Colloid Interf. Sci.,* Effects of a Magnetic Field on the Formation of $CaCO_3$ Particles **156,** 90- 95 (1993).
[28]L. Pach, S. Duncan, R. Roy, and S. Komarneni, Effects of a Magnetic Field on precipitation of $CaCO_3$, *J. Materials Sci. Letters* **15**, 613-615 (1996).
[29]G. Paskalov, Activated water apparatus and methods, US Pat # US 7,291,314 B2, Nov 2007.
[30]R. Roy, M. L. Rao and J. Kanzius, Observations of polarized RF radiation catalysis of dissociation of H2O-NaCl solutions, *Mater. Res. Innov.* **12**, 3-6 (2008).
[31]M.L. Rao, R. Roy and S. Sedlmayr: Single Mode 2.45 GHz Microwave, and Polarized 13.56 MHz Radiofrequency Radiation effects on the Structure and Stability of Liquid Water, *Proc. MRS 2007 Fall Meeting, Boston, USA,* November (2007).

PREPARATION OF $CuInS_2$ FILMS BY ELECTRODEPOSITION: EFFECT OF METAL ELEMENT ADDITION TO ELECTROLYTE BATH

Tomoya Honjo, Masayoshi Uno and Shinsuke Yamanaka
Division of Sustainable Energy and Environmental Engineering, Graduate School of Engineering, Osaka University
2-1 Yamadaoka, Suita, Osaka 565-0871, Japan

ABSTRACT

Gallium incorporation in $CuInS_2$ films were successfully prepared by electrodeposition. The electrodeposition was carried out in a three-electrode system, using SUS316 plate as a working electrode, Pt mesh as a counter electrode and Ag/AgCl (saturated KCl) electrode as a reference electrode. The solution for deposition contained desirable concentrations of Cu^{2+}, In^{3+}, Ga^{3+}, $S_2O_3^{2-}$ and LiCl, and the pH value was adjusted with a buffer solution of pH=3. Deposition was performed potentiostatically. After the electrodeposition, the films were annealed without H_2S gas. The limiting current was observed between 0.8 and 0.9 V vs. Ag/AgCl, which shifted towards higher potentials compared with the case of Ga^{3+} ion-free solution. Deposited films had p-type conductivity. With increasing the gallium content in the film, its band gap was varied between 1.6 and 1.7 eV, and the conduction band edge of the compounds increased significantly while the valence band edge remained at nearly the same energy position.

INTRODUCTION

Ternary semiconductor compounds of I-III-VI_2 type have been widely studied for use in optical devices and solar cells. Above all, $CuInS_2$ is suitable for use in solar cells. The compound has a direct band gap energy of 1.5 eV and high light absorption coefficient. Furthermore, it is composed of non-toxic elements. Its band gap energy can be controlled by doping gallium[1], and its conduction type can be also changed by adjusting the ratio of elements.[2,3] Efficiency of 12.2 % has been obtained using $CuInS_2$ thin film based solar cells.[4] The theoretical efficiency calculated for homojunction structures using this materials is about 32 %.[5] It is expected to be a substitute for silicon solar cells.

$CuInS_2$ thin films have been deposited using various techniques, such as physical evaporation,[6] rapid thermal process,[7] chemical vapor deposition,[8] spray pyrolysis,[9] chemical bath deposition[10] and electrodeposition.[11-13] Of special note, electrodeposition is low cost process and simple technique. Moreover, this technique is suitable for growing large area which can be prepared at a low temperature. These features indicate the advantage for the industrialization of manufacturing $CuInS_2$ solar cells.

Many authors have studied the electrodeposition of $CuInS_2$ using electrodeposited Cu-In precursors followed by annealing in H_2S atmosphere.[11] Yukawa et al. have reported electrodeposited $CuInS_2$ films by using $Na_2S_2O_3$ without H_2S gas.[12] It is a disadvantage to use H_2S gas because it is toxic. Martinez et al. have reported electrodeposition of Cu-In-S in a buffer solution followed by annealing in a N_2 atmosphere.[13] Although some authors reported depositing $CuInS_2$ thin films by electrodeposition technique, there are few studies of doping elements into $CuInS_2$ films by this technique. In this study, Ga incorporation in $CuInS_2$ films were fabricated with electrodeposition technique in order to control the band gap energy by adding Ga^{3+} ion into the solution, and then some physical characteristics of the films were measured.

EXPERIMENTAL

The solution for deposition contained 12.5 mM Cu^{2+}($CuCl_2 \cdot 2H_2O$), 12.5 mM In^{3+}($InCl_3$), 125 mM $S_2O_3^{2-}$($Na_2S_2O_3 \cdot 5H_2O$), 250 mM LiCl and desirable concentrations of Ga^{3+}($GaCl_3$) with a buffer solution of pH=3. The concentrations of Ga^{3+} are shown in Table I. The pH value of the solution was adjusted to the required value by adding HCl or NaOH (shown in Table I). The electrodeposition was carried out in a three-electrode system, using SUS316 plate as a working electrode, Pt mesh as a counter electrode and Ag/AgCl (saturated KCl) electrode as a reference electrode. Deposition was performed potentiostatically. The applied potential is shown in Table I. SUS316 plate was polished with sand paper and ultrasonicated in acetone for 1 min, and then immersed in 1.2 M HCl for 1 min. The washed plate was anodically-electrolyzed in 1 M NaOH solution at a current density of 7.5 mA/cm^2 for 2 min. The duration of deposition was 1500 s. After the electrodeposition, the films were annealed at 623 K (A-0) or 673 K (A-0.5, A-1.0 and A-1.5) in a vacuum for 1 h in order to increase the crystallinity. Finally, in order to remove Cu-In compounds such as Cu_7In_3, the samples were etched by scanning potential from -0.5 to +0.5 V at the scan rate 5 mV/s in 0.1 M H_2SO_4 solution, and potentiostatically at -0.3 V in the solution for 1 min.[14]

The crystal structure was characterized by X-ray diffraction (XRD) (RIGAKU RINT2000) using Cu-Ka radiation. The observed phases were determined by comparing the d-spacing with JCPDS data files. The surface morphology was evaluated by a field-emission scanning electron microscope (FE-SEM) (JEOL JSM-6500F). The atomic composition of thin films was investigated by EDX (HORIBA EX-200).

The preparation of the electrodes for electrochemical studies was as follows: the electric contacts of $CuInS_2$ thin films were done by a copper wire, which was attached to the film with conductive Pt paste. The body of the wire was encased in silicon tubing and insulated using an epoxy coating that also covered the sample edges. The total area expose of the electrode was approximately 1 cm^2.

The current-voltage measurement was carried out using a potentiostat (Solartron SI1260) and a 500 W Xenon short arc lamp (USHIO SX-UID500X). The band gap energy of the film was determined by using the Xenon lamp, a monochromator and optical filters. Both electrochemical measurements were carried out in aqueous solutions containing 0.1 M Na_2SO_4.

Table I. Conditions of the electrodeposition.

Label	Ga^{3+} (mM)	Ga^{3+}/In^{3+}	pH	Applied potential (V vs. Ag/AgCl)
A-0	0	0	3.2	-0.95
A-0.5	6.25	0.5	2.4	-0.95
A-1.0	12.5	1.0	2.0	-0.85
A-1.5	18.75	1.5	2.0	-0.85

RESULTS AND DISCUSSION

The electrodeposition of Cu-In-S thin films on cathodes is most likely caused as follows:[13-15]

$$Cu^{2+} + 2e^{-} \rightarrow Cu \qquad E_0(SHE) = 0.337(V)$$

$$In^{3+} + 3e^{-} \rightarrow In \qquad E_0(SHE) = -0.342(V)$$

$$Ga^{3+} + 3e^{-} \rightarrow Ga \qquad E_0(SHE) = -0.529(V)$$

$$S_2O_3^{2-} + H^{+} \rightarrow S + HSO_3^{-}$$

$$iCu + jIn + kGa + lS \rightarrow Cu_iIn_jGa_kS_l$$

The pH of the solution has an important role as it affects the deposition of sulfur. As a result, the pH controls the formation of soluble complexes of copper. In addition, the higher pH of the solution containing Ga^{3+} ion caused hydrolysis. Considering this, lower pH was adopted for this solution.

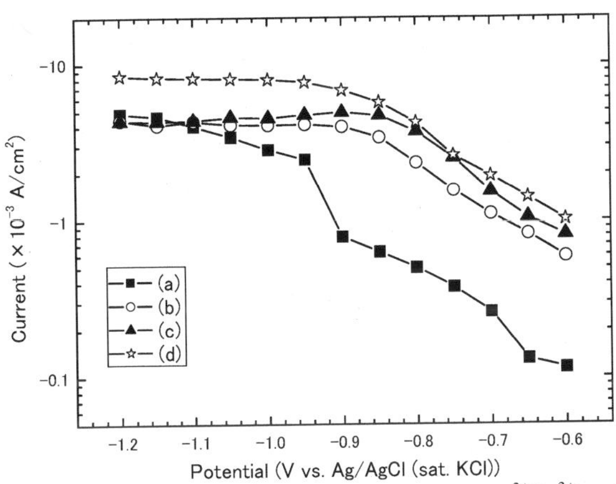

Figure 1. Polarization curve of solutions with the ratio Ga^{3+}/In^{3+} = (a) 0 (pH=3.2), (b) 0.5 (pH=3.2), (c) 1.0 (pH=2.8), (d) 1.5 (pH=2.4).

Figure 1 shows a polarization curve of the solutions containing various amount of Ga^{3+} ion. In the case of the Ga^{3+} ion-free solution, the current decreased rapidly at potential between -0.75 and -0.85 V. A limiting current was observed around -0.95 V. This indicates that it is favorable to deposit around -0.95 V in the Ga^{3+} ion-free solution. Meanwhile in the case of the solution containing Ga^{3+} ion, a magnitude of the current was larger than in the case of the Ga^{3+} ion-free solution, and it decreased slightly. It may be associated with the reduction of Ga^{3+} ion. The limiting current was observed between 0.8 and 0.9 V vs. Ag/AgCl, which shifted towards higher potentials compared with the case of the Ga^{3+} ion-free solution. This result agrees with the tendency measured by Bouabid et al.[16]

According to Figure 1, the potentials for electrodeposition of the films were selected -0.95 V for A-0 and A-0.5 in the solution, and -0.85 V for A-1.0 and A-1.5.

Table II. Lattice constant and atomic composition of Cu, In, S and Ga for films.

Label	Lattice constant (Å)		Atomic composition (at.%)				Ga/(In+Ga) (%)
	a	c	S	Cu	Ga	In	
A-0	5.526	11.11	43.7	28.7	-	27.6	-
A-0.5	5.507	11.07	44.4	29.4	1.4	24.8	5.3
A-1.0	5.502	11.04	42.6	28.8	2.0	26.4	7.1
A-1.5	5.484	10.99	44.6	27.8	3.7	23.9	13.4

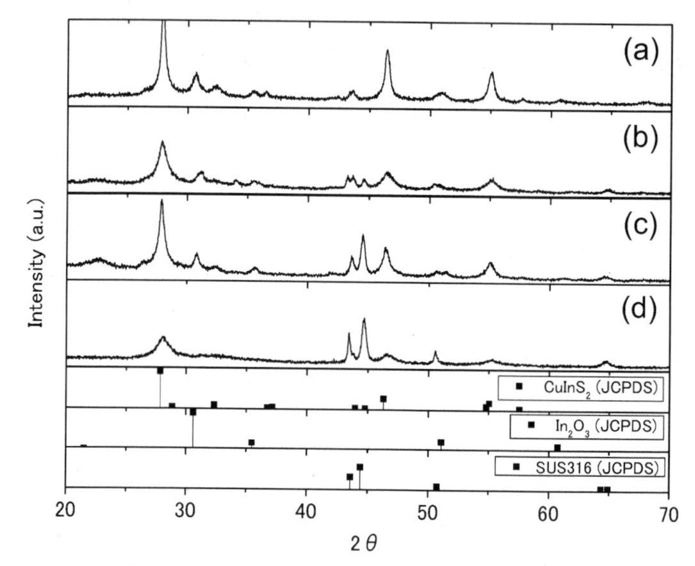

Figure 2. XRD patterns for annealed samples of films: (a) A-0, (b) A-0.5, (c) A-1.0, (d) A-1.5.

The XRD patterns for annealed samples of films are shown in Figure 2. All samples showed the peak from chalcopyrite phase. As the concentration of Ga^{3+} in the solution was increased, the peak gradually decreased. This may be attributed to the incorporation of gallium in the Indium sites which distort the normal lattice structure of $CuInS_2$. Also, In_2O_3 phase was identified in the sample of A-0, A-0.5 and A-1.0. Each lattice constant of the film is shown in Table II. The lattice constant of the film obtained from the solution containing Ga^{3+} ion was shorter than that of $CuInS_2$ without gallium. This result indicates that gallium was taken in $CuInS_2$ successfully.

The result of EDX analysis is also shown in Table II. As the concentration of Ga^{3+} ion in the electrolytic solution increased, the Ga/(In+Ga) ratio of the film as well as the amount of electrodepositing gallium increased. However, Ga/(In+Ga) ratio of the film was lower than Ga^{3+}/In^{3+} in the solution. This is probably because the reduction of In^{3+} ion occurs more easily than that of Ga^{3+} ion and more and more indium was electrodeposited. Cu:In:S ratio of all samples from EDX analysis were not exactly 1:1:2. This indicates that some impurities affected the analysis.

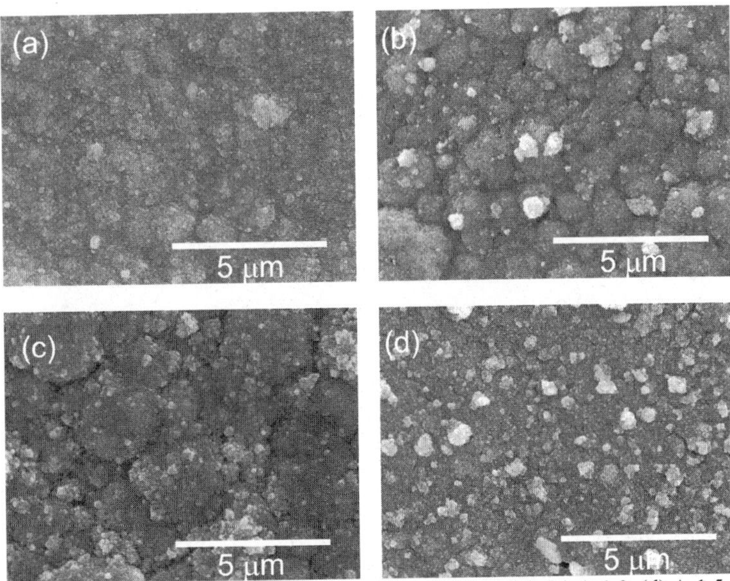

Figure 3. SEM micrograph of the films: (a) A-0, (b) A-0.5, (c) A-1.0, (d) A-1.5.

Figure 3 shows SEM micrograph of the films. A-0 had relatively flat and smooth surface. The surfaces of the film containing gallium were not powdery but rougher than that of the film without gallium, and they had uniform morphology despite the difference in the concentration of Ga^{3+} ion in the solution.

The results of the current-voltage measurement of the films shown in Figure 4 indicate that all the films behave as p-type semiconductors. In dark conditions, the sample showed a rectifying property that anodic current flowed under anodic polarization but vary low cathodic current under cathodic polarization. By the illumination of light, cathodic photocurrent was observed at negative potential, but it was not well pronounced. Sebastian et al. also reported the same phenomena in electrodeposited $CuInSe_2$ films and described that it may be attributed to mixed conductivity owing to the contribution from superficial n-type as well as bulk p-type conductivity.[17]

The applied potential at which photocurrent becomes zero, flat band potential (U_{fb}), could be estimated from these polarization curves. The value of flat band potential of each sample shown in Table III is relatively constant.

Table III. Electrical and optical properties of the samples.

Label	Conduction type	U_{fb} (V vs. SHE)	Band gap energy (eV)
A-0	p	0.32	1.52
A-0.5	p	0.26	1.60
A-1.0	p	0.31	1.65
A-1.5	p	0.32	1.70

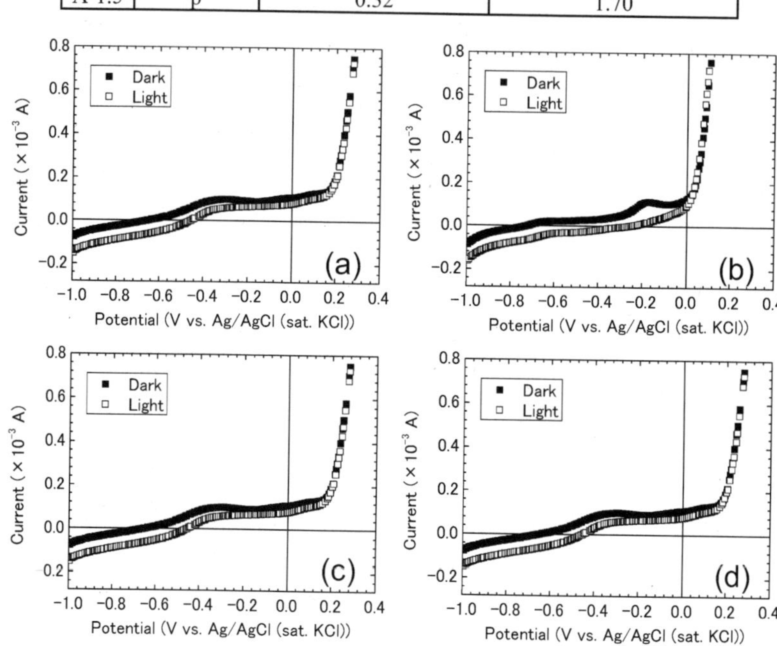

Figure 4. Current-voltage measurement of the films: (a) A-0, (b) A-0.5, (c) A-1.0, (d) A-1.5.

Photocurrent action spectrum of each specimen as a function of the photon energy, $h\nu$, is plotted in Figure 5. The value of quantum efficiency, Φ, in Figure 5 was calculated using:

$$\Phi = \frac{I_p h\nu}{Pe}$$

where P is the illuminating light power and e is the charge of an electron, 1.6×10^{-19} C. During the photocurrent action spectrum measurement, the applied potential of each specimen was held at the value -0.3 V vs. Ag/AgCl.

The absorption coefficient, α of a crystalline material depends on the photon energy, $h\nu$, according to the following equation:

$$\alpha = A\frac{(h\nu - E_g)^{\frac{n}{2}}}{h\nu}$$

where E_g is the semiconductor band gap energy, A is a constant and n is an integer which depends on whether the electron excitation is direct (n=1) or indirect (n=4). On the assumption that Φ is proportional to α, the type of transition and the band gap energy can be determined from the plot of the following equation:

$$(\Phi h\nu)^{2/n} = A'(h\nu - E_g)$$

where A' is a constant.[18] Figure 5 shows plots of $(\Phi h\nu)^{1/2}$ vs. $h\nu$. A good linear relationship was obtained near the band gap energy for each specimen in the hypothesis of direct optical transitions. Plots with n=4 did not yield straight lines for all films. This indicates that the excitation process of all the films was a direct transition. From the extrapolation of the plots near the band gap, the band gap energy for the Ga-free film was estimated to be 1.52 eV. This value agrees well with previously reported band gap energy for $CuInS_2$ films, 1.4-1.5 eV.

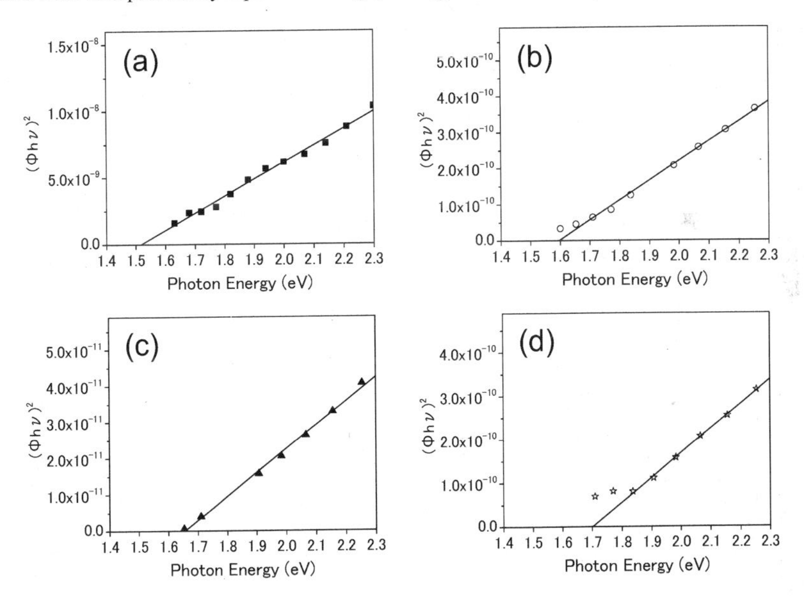

Figure 5. Plots of $(\Phi h\nu)^{1/2}$ versus photon energy for the films: (a) A-0, (b) A-0.5, (c) A-1.0, (d) A-1.5.

The value of band gap energy of the films was proportional to the Ga^{3+}/In^{3+} ratio in the electrolytic solution. Maximum value of band gap energy was 1.70 eV when the Ga^{3+}/In^{3+} ratio was 1.5. Wei et al. said that the difference between the electronegativity of In^{3+} and Ga^{3+} ions can explain the increase in the band gap when indium is substituted by gallium in the CIS film.[19] This result also indicate that gallium incorporation in $CuInS_2$ films were successfully prepared.

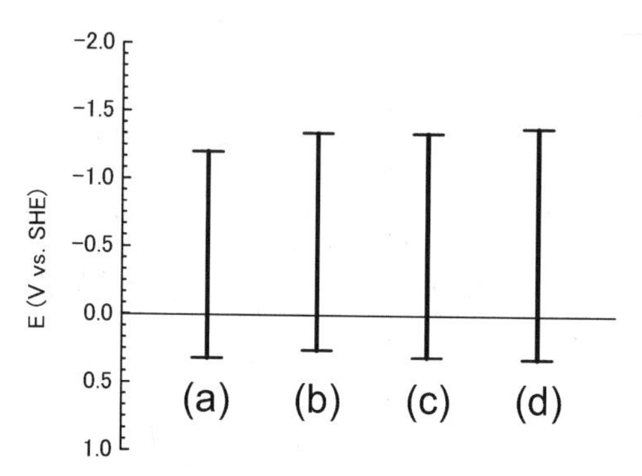

Figure 6. Energy level diagram of the films: (a) A-0, (b) A-0.5, (c) A-1.0, (d) A-1.5.

The flat band potential and band gap values were used to calculate the conduction band edges of the films. For p-type semiconductor, flat band potential is almost equal to the valence band E_{VB}. Thus, the conduction band E_{CB} can be calculated from following equations.

$$E_{CB} = E_{VB} - E_g = U_{fb} - E_G$$

Figure 6 shows energy level diagram of the films. This figure shows that any of the valence band edge of the films were located at relatively the same energy position (= 0.3 V vs. SHE). On the contrary, conduction band edge shifted to higher energy as Ga^{3+}/In^{3+} ratio increased.

CONCLUSION

Gallium incorporation in $CuInS_2$ films were successfully prepared by electrodeposition. The films were sintered without H_2S. A limiting current was observed between 0.8 and 0.9 V vs. Ag/AgCl, which shifted towards higher potentials compared with the case of Ga^{3+} ion-free solution. As Ga^{3+}/In^{3+} ratio in the electrolytic solution increased, lattice constants of the films were decreased and Ga/In ratios of the film were increased. Deposited films had p-type conductivity. With increasing the gallium content in the film, its band gap was varied between 1.6 and 1.7 eV, and the conduction band edge of the compounds increased significantly while the valence band edge remained at approximately the same energy position.

REFERENCES

[1] R. Kaigawa, A. Ohyama, T. Wada, R. Klenk, Electrical properties of homogeneous $Cu(In,Ga)S_2$ films with varied gallium content, *Thin Solid Films*, **515**, 6260-6264 (2007).

[2] B. Tell, J.L. Shay, Room-Temperature Electrical Properties of Ten I-III-VI_2 Semiconductors, *J. Appl. Phys.*, **43**, 2469-2470 (1972).
[3] H. J. Hsu, M. H. Yang, R. S. Tang, T. M. Hsu, H. L. Hwang, A novel method to grow large $CuInS_2$ single crystals, *J. Cryst. Growth*, **70**, 427-432 (1984).
[4] D. Braunger, Th. Duerr, D. Hariskos, C.h. Koeble, Th. Walter, N. Wieser, H.W. Schock, Improved open circuit voltage in $CuInS_2$-based solar cells, *Conference Record of the IEEE Photovoltaic Specialists Conference*, 1001-1004 (1996).
[5] J.M. Meese, J.C. Manthuruthil, D.R. Locker, $CuInS_2$ diodes for solar energy conversion, *Bull. Am. Phys. Soc.*, **20**, 696 (1975).
[6] M. Kanzari, B. Rezig, Effect of deposition temperature on the optical and structural properties of as-deposited $CuInS_2$ films, *Semicond. Sci. Technol.*, **15**, 335-340 (2000).
[7] K.Siemer, J.Klaer, I.Luck, J.Bruns, R.Klenk, D.Braeunig, Efficient $CuInS_2$ solar cells from a rapid thermal process (RTP), *Sol. Energy Mater. Sol. Cells*, **67**, 159-166 (2001).
[8] J.H. Park, M. Afzaal, M. Kemmler, P. O'Brien, D.J. Otway, J. Raftery, J. Waters, The deposition of thin films of CuME by CVD techniques (M = In, Ga and E = S, Se), *J. Mater. Chem.*, **13**, 1942-1949 (2003).
[9] M. Krunks, O. Bijakina, T. Varema, V. Mikli, E. Mellikov, Structural and optical properties of sprayed $CuInS_2$ films, *Thin Solid Films*, **338**, 125-130 (1999).
[10] S. Bini, K. Bindu, M. Lakshmi, C. Sudha Kartha, K.P. Vijayakumar, Y. Kashiwaba, T. Abe, Preparation of $CuInS_2$ thin films using CBD CuS films, *Renew. Energy*, **20**, 405-413 (2000).
[11] R.P. Wijesundera, W. Siripala, Preparation of $CuInS_2$ thin films by electrodeposition and sulphurisation for applications in solar cells, *Sol. Energy Mater. Sol. Cells*, **81**, 147-154 (2004).
[12] T. Yukawa, K. Kuwabara, K. Koumoto, Electrodeposition of $CuInS_2$ from aqueous solution (II) electrodeposition of $CuInS_2$ film, *Thin Solid Films*, **286**, 151-153 (1996).
[13] A.M. Martinez, A.M. Fernandez, L.G. Arriaga, U. Cano, Preparation and characterization of Cu-In-S thin films by electrodeposition, *Materials Chemistry and Physics*, **95**, 270-274 (2006).
[14] J. Kois, S. Bereznev, O. Volobujeva, E. Mellikov, Electrochemical etching of copper indium diselenide surface, *Thin Solid Films*, **515**, 5871-5875 (2007).
[15] R.N. Bhattacharya, J.F. Hiltner, W. Batchelor, M.A. Contreras, R.N. Noufi, J.R. Sites, 15.4% $CuIn_{1-x}Ga_xSe_2$-based photovoltaic cells from solution-based precursor films, *Thin Solid Films*, **361**, 396-399 (2000).
[16] K. Bouabid, A. Ihlal, A. Manar, A. Outzourhit, E.L. Ameziane, Effect of deposition and annealing parameters on the properties of electrodeposited $CuIn_{1-x}Ga_xSe_2$ thin films, *Thin Solid Films*, **488**, 62–67 (2005).
[17] P.J. Sebastian, M.E. Calixto, R.N. Bhattacharya, Rommel Noufi, CIS and CIGS based photovoltaic structures developed from electrodeposited precursors, *Sol. Energy Mater. Sol. Cells*, **59**, 125-135 (1999).
[18] J. W. Halley, M. Kozlowski, M. Michalewicz, W. Smyrl and N. Tit, Photoelectrochemical spectroscopy studies of titanium dioxide surfaces: theory and experiment, *Surf. Sci.*, **256**, 397-408 (1991).
[19] S.H. Wei, S.B. Zhang, A. Zunger, Effects of Ga addition to CuInSe2 on its electronic, structural, and defect properties, *Appl. Phys. Lett.*, **72**, 3199-3201 (1998).

PREPARATION OF HIGH-J_c MOD-YBCO FILMS FOR FAULT CURRENT LIMITERS

M. Sohma, W. Kondo, K. Tsukada, I. Yamaguchi, T. Kumagai, T. Manabe, K. Arai, H. Yamasaki

National Institute of Advanced Science and Technology (AIST), AIST Tsukuba Central 5, 1-1-1 Higashi, Tsukuba, Ibaraki 305-8565, Japan

ABSTRACT

A large-area, 210 mm × 30 mm, $YBa_2Cu_3O_7$ (YBCO) superconducting film with high superconducting critical current density (J_c) was prepared on a CeO_2-buffered sapphire substrate by a metalorganic deposition method. The mapping- J_c values reached as high as a 4.0-5.7 MA/cm^2 range at 77.3 K by an inductive method. A fault current module was produced using the YBCO film (effective area: 60 mm × 28 mm) with an Au-Ag shunt alloy layer, and a set of an external resistance and a condenser. Switching experiments were carried out by applying sinusoidal over-currents with an over-current duration of 100 ms, i.e., for 5 cycles at 50 Hz. A peak voltage of 245 V along the effective length of 60 mm was achieved. This voltage corresponds to a high electric field of 40 V/cm.

INTRODUCTION

Metalorganic deposition (MOD) is one of the most promising methods to prepare various functional oxide films, since the method has an advantage of precise composition control and potential for manufacturing complex-metal-oxide films such as high-temperature superconducting (HTS) $YBa_2Cu_3O_7$ (YBCO) [1-3]. This technique can be applied to large-area HTS films at low cost necessary for a resistive-type fault current limiter (FCL) which limits the prospective fault current when a fault occurs.

For this application, YBCO films are required to form on large-area sapphire (α-Al_2O_3) substrates; large-area films are needed because at normal operation the rated current should flow without resistance through the FCL and, when a fault occurs, it should withstand high electric fields; and sapphire substrates are preferred since their excellent thermal conductivity and mechanical strength prevent breakage due to large thermal stress upon quenching [4]. However, direct growth of YBCO on sapphire is very difficult, thus, a proper buffer layer such as CeO_2 is necessary between YBCO and sapphire to prevent chemical reaction at the interface as well as to relax the large lattice mismatch between them, i.e., ~10%. In our previous articles, we showed that vacuum-evaporated CeO_2 buffer layers having high (100)-orientation and very smooth surface are quite suitable for preparing YBCO films by MOD [2] and that a *c*-axis oriented YBCO film grown on a 300 mm × 100 mm CeO_2-buffered sapphire substrate demonstrated high inductive- critical-current-density (J_c) values with an average 2.6 MA/cm^2 at 77K [1]. In the MOD-YBCO procedure a fluorine-free metal acetylacetonate-based coating solution including stoichiometric ingredients is spin-coated onto the CeO_2- buffered sapphire substrates. After the coating the organic components decompose to amorphous material by prefiring and then, by high-temperature heat treatment under precisely controlled conditions, to the crystallized YBCO phase.

This paper presents the preparation of a high-J_c MOD-YBCO film having an area of 210 mm × 30 mm and the results of fundamental over-current tests for a fault current module using the film. The superconducting film element in the module is covered with Au-Ag alloy shunt layers, which we have proposed to prevent the "hot-spots" problem. We have achieved high electric fields using the shunt [5-7].

EXPERIMENTAL

Preparation of YBCO Films

A large-area (210 mm × 30 mm) polished R-cut sapphire substrate (1 mm-thick., Kyocera Corp., SA100, Japan) was annealed at 1000°C for 1 hour in an electric furnace to promote atomically flat terrace of the surface as reported in our previous study [1]. A CeO_2 buffer layer (24 nm) was deposited on the substrate using an electron-beam deposition apparatus. In our MOD process, a coating solution was made from fluorine-free metal acetylacetonates (Nihon Kagaku Sangyo Co., Ltd., Y:Ba:Cu = 1:2:3 molar ratio), and was spin-coated onto the CeO_2-buffered substrate. After the coating the organic components decompose to amorphous material by prefiring at 500°C and then, by high-temperature heat treatment at around 750°C under the control of oxygen partial pressure, to the crystallized YBCO phase. The precise preparation conditions of the CeO_2 buffer and of the MOD-YBCO layer have been reported elsewhere [2,3]. An Au-Ag alloy layer (60 nm) was deposited on the YBCO by a sputtering process using an Au-23wt% Ag alloy target. The Au-Ag/ YBCO/ CeO_2/ sapphire composite was divided into two to be adapted for the limit of the present AC power supply in the over-current test. Silver electrodes were also deposited at both ends of the sample by sputtering. The effective area of the superconducting film element was 28 mm-width × 60 mm-length. A standard x-ray diffraction system (XRD: MAC Science MFX-HP which can accommodate a large size sample) was employed to examine the crystalline quality for the YBCO layer. Spatial distribution of superconducting J_c for the YBCO film was obtained by mapping analysis (at intervals of 5 mm) of inductive- J_c data (AIST and Hayama Inc., J_c-n measuring system; probe coil diameter: 5 mm) at 77.3 K through whole area of the specimen except for edge parts.

Over-Current Tests

A schematically illustrated diagram (figure 1) shows the current limiting module used for the over-current tests. An external resistance R, a noninductively wound manganin wire (0.33 Ω), and a condenser (120 μF) were connected in parallel to the film element to protect the superconducting films from the "hot-spots" problem during quenching, and they were immersed in liquid nitrogen. Four voltage taps 1 to 4 were attached to the film element. The space between the voltage taps, 1-2, 2-3 and 3-4 was 20 mm each. Switching experiments were carried out by applying sinusoidal over-currents to the parallel-connected specimen and the external resistance, with an over-current duration of 100 ms, i.e., for 5 cycles at 50 Hz. The AC power supply was operated in the voltage-control mode. In this mode, the power supply induces the output voltage V_{power} which is proportional to the voltage applied to the power supply, where a resistance of 0.2 Ω was connected to the power supply.

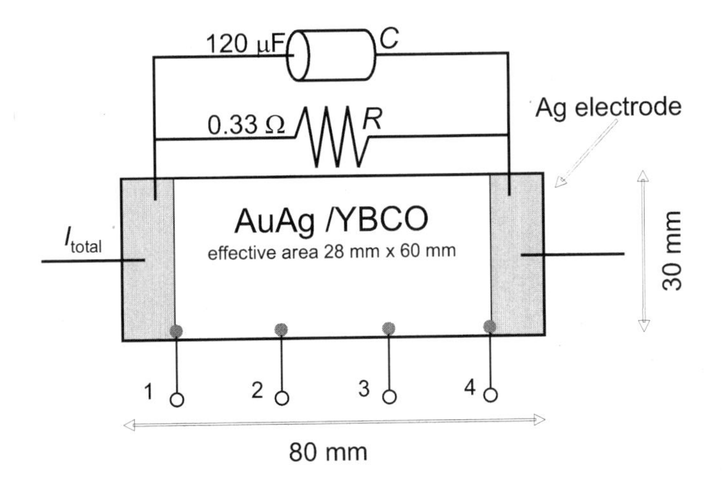

Figure 1 Schematic illustration of a current limiting module which is composed of a superconducting film element, an external resistance R (a noninductively wound manganin wire), and a condenser C. They are connected in parallel each other

RESULTS AND DISCUSSION

Superconducting Property of YBCO Film

A typical XRD θ-2θ pattern of the 140-nm-thick YBCO film is exhibited in figure 2 on a logarithmic scale. The completely *c*-axis-oriented YBCO phase was observed, and the *a*-axis-oriented one was not detected. In the XRD patterns measured at various positions of the large-area film there was no significant difference. Impurity phases, such as $BaCeO_3$, were not observed except for a small amount of the Y124 phase as shown in the figure. The existence of the small amount of the Y124 phase, which is often observed in cross sectional transmission electron microscopic images, does not always prevent the YBCO films from having a high-J_c property [8].

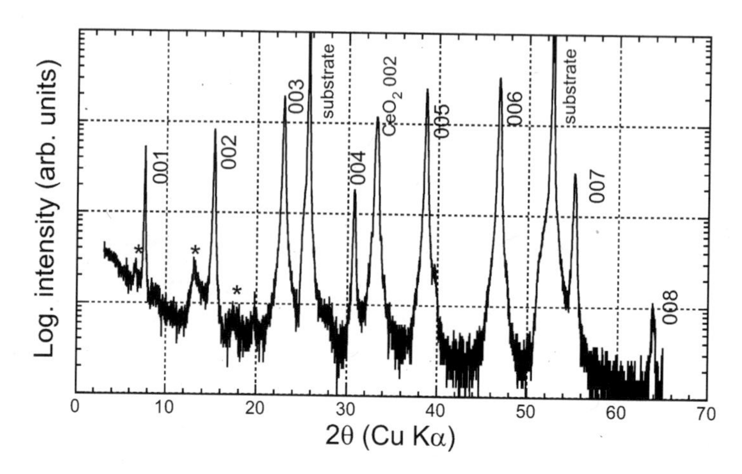

Figure 2 A typical XRD pattern of the c-axis oriented YBCO film on a CeO_2-bufffered sapphire substrate. Asterisks indicate the diffraction from the Y124 impurity phase

The J_c distribution map of the YBCO film measured by an inductive method at 77.3 K is presented in figure 3. We did not measure the edge parts of the film because when the probe coil crosses the film edges, the third harmonic signal versus coil current curves in the inductive measurement became spurious [9,10]. The J_c values reached as high as a 4.0-5.7 MA/cm^2 range. The shaded areas indicate the position of silver electrodes which were deposited after the J_c measurements.

	4.46	4.55	4.63	4.74	4.89	4.87	4.79	4.77	4.82	4.86	5.13	5.04	4.95	2.88	4.80
	4.02	4.10	4.04	4.12	4.32	4.49	4.41	4.41	4.61	4.85	4.98	4.95	NA	5.11	3.68
	4.40	4.56	4.50	4.42	4.42	4.68	4.69	5.03	5.08	5.01	4.88	4.67	4.77	4.68	2.11
	5.07	4.95	4.76	4.60	5.06	5.18	4.85	4.60	5.72	5.62	5.53	5.42	4.95	5.00	3.24

30 mm

80 mm

Figure 3 J_c mapping of the YBCO film (30 mm x 80 mm) on the CeO_2-bufffered sapphire substrate. Shaded areas were covered with Ag electrodes after the measurements

Fault Current Limiting Property

Superconducting critical current (I_c) of the film element was measured to be 240 A. No quenching was observed when the peak current was 275 A for 5 cycles. The test results of the over-current test using the current limiting module in figure 1 are shown in figures 4(a) and 4(b). Figure 4(a) shows the initial part of the quenching, in which V_{film}, V_{12}, V_{23}, V_{34} refer to the total voltage of the film element, the output voltages between voltage taps, 1-2, 2-3 and 3-4, respectively. Before the fault current flows all the voltages kept at zero. According to the behavior of the voltage curves, quenching was probed to initiate at time (t) = 120.9 ms between voltage taps 3 and 4 (V_{34}), thereafter V_{23} (t = 123.2 ms) and V_{12} (t = 133.2 ms) started to rise in this order, corresponding to the expansion of the normal zone. The V_{film} represents the total voltage along the effective length of the film element. The normal zone propagation velocity was calculated to be approximately 5 m/s for one direction.

Moreover, the behaviors of the total current (I_{total}) and the film current (I_{film}) are shown in figure 4(b) with V_{film} and V_{power}. Peak voltage (245 V) of V_{film} along the effective length of 60 mm was achieved at t = 135.1 ms. This voltage corresponds to a high electric field of 40 V/cm. Without the fault current module I_{total} would reach approximately 2.5 kA. By using our module, I_{total} was limited to as small as 815 A at the initial peak immediately after quenching, and this I_{total} value was comparable to the 3 × I_c, 720A.

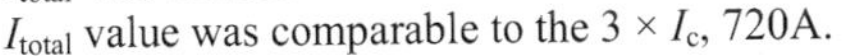

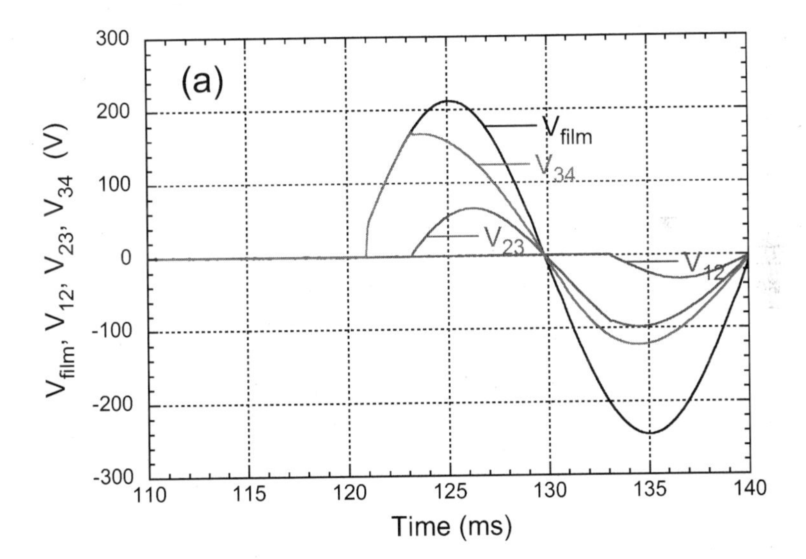

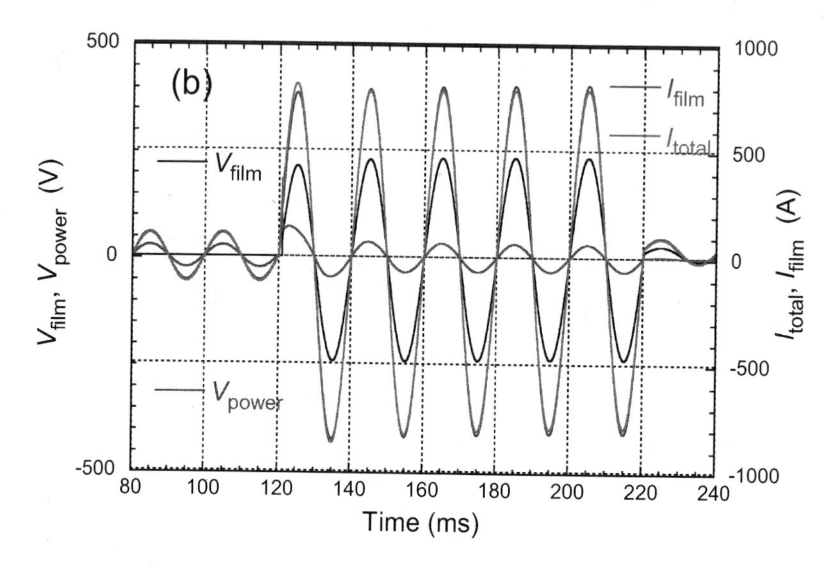

Figure 4 Results of the over-current test using the fault current device: (a) voltages in film elements in the initial part of the quenching, (b)behaviors of the total current, the film current, the film voltage, and voltage of the power source

REFERENCES

[1] T. Manabe, M. Sohma, I. Yamaguchi, W. Kondo, K. Tsukada, S. Mizuta, and T. Kumagai, *Physica C*, Vol 412-414, 2004, p 896-899

[2] M. Sohma, I. Yamaguchi, K. Tsukada, W. Kondo, K. Kamiya, S. Mizuta, T. Manabe and T. Kumagai, *IEEE Transactions on Appl. Supercond.*, Vol 15, 2005, p 2699-2702

[3] T. Manabe, M. Sohma, I. Yamaguchi, W. Kondo, K. Tsukada, K. Kamiya, S. Mizuta, and T. Kumagai, *IEEE Transactions on Appl. Supercond.*, Vol 15, 2005, p 2923-2926

[4] K. Shimohata, S. Yokoyama, T. Inaguchi, S. Nakamura, Y. Ozawa, *Physica C*, Vol 372-376, 2002, p 1643-1648

[5] H. Yamasaki, M. Furuse, and Y. Nakagawa, *Appl. Phys. Lett.*, Vol, 85, 2004, p 4427-4429

[6] K. Arai, H. Yamasaki, K. Kaiho, M. Furuse, and Y. Nakagawa, *IEEE Trans. Appl. Supercond.*, Vol 17, 2007, p 1843-1846

[7] M. Furuse, H. Yamasaki, T. Manabe, M. Sohma, W. Kondo, I. Yamaguchi, T. Kumagai, K. Kaiho, K. Arai, and Y. Nakagawa, *IEEE Trans. Appl. Supercond.*, Vol 17, 2007, p 3479-3482

[8] M. Sohma, K. Tsukada, I. Yamaguchi, K. Kamiya, W. Kondo, S. Mizuta, T. Manabe, T. Kumagai, *J. Phys.: Conferences Series*, Vol 43, 2006, p 349-352

[9] J. H. Claasen, M. E. Reeves, and J. Soulen Jr., *Rev. Sci. Instrum.*, Vol 62, 1991, p 996-1004

[10] Y. Mawatari, H. Yamasaki, and Y. Nakagawa, *Appl. Phys. Lett.*, Vol 81, 2002, p 2424-2426

Nanotechnology for Power Generation

MODELING OF ELECTROMAGNETIC WAVE PROPAGATION OF NANO-STRUCTURED FIBERS FOR SENSOR APPLICATIONS

Neal T. Pfeiffenberger and Gary R. Pickrell
Virginia Tech,
Blacksburg, VA, USA

ABSTRACT

The focus of this work is centered on understanding the modal characteristics at optical frequencies for the electromagnetic wave propagation in nano-structured holey fibers. The fiber optic structures simulated in this work were created using the Comsol Multiphysics software package. An SEM cross-section of a random-hole fiber image was imported into Comsol and the results of the modeling have been shown. This process allows for optimization of a fiber whose structure is not easily replicated in a FEM environment.

INTRODUCTION

Design of the appropriate fiber structure can be very important in today's sensor and data transmission markets. Optical fibers are immune to electromagnetic interference, which greatly reduces the noise as compared to conventional electrical components which can be very important in some applications. Many articles have been published on the use of optical fibers for sensors in a wide variety of applications including defense, civil infrastructure monitoring, down-hole oil well sensing, etc.

The fibers under study in this work are called random hole optical fibers (RHOF). The name originates from the fact that the "holes" in the fiber are randomly located in the spatial domain, and can be controlled to sizes in the micron to nanometer size range by controlling the conditions of fabrication. They are a new type of fiber structure in contrast to the ordered holes present in the microstructure optical fiber (MOF) or the photonic crystal fiber (which is a subset of the MOFs). The RHOFs use thousands of air holes to confine the light to the core of the fiber through what is believed to be an average index guiding mechanism. One of the important aspects of the air holes is the fact that they can allow access to the interior of the fiber while still maintaining the modal properties of the fiber[1]. This can be very important in sensing applications. In addition, since the structures can be produced in pure silica (no dopants required), the high temperature capabilities are improved relative to conventional doped fibers. Current telecommunication fibers incorporate dopants into the silica glass structure to change the refractive index of the silica. These doped regions are not robust in high temperature applications because their dopants can diffuse from the core to the cladding layers or vice versa, which leads to changes in their modal properties.

The continual work on the characterization and fabrication of ordered hole fibers can be seen throughout the literature[2-6], beginning with the work of Knight, Birks, Russell and Atkin. Ordered hole fibers are oriented in a preform which is comprised of an array of stacked longitudinal tubes. These longitudinal tubes are then carefully drawn to produce an ordered arrangement of holes in the finished fiber.

In contrast to the ordered hole fibers, the random hole fibers are created by producing holes *in situ* during the fiber drawing stage. This produces a vast number of "holes", which can be controlled to have diameters that range from a few nm to the µm size range, by controlling the conditions of the fiber drawing process and the preform characteristics. Like the ordered hole fibers, RHOF's are composed of longitudinal holes, which run parallel to the fiber axis. The large difference between the refractive index of air and silica glass along with the numerous possibilities regarding the arrangement of these air holes may allow RHOF's to be used in a wide range of applications.

This study's main focus is on the modeling of these structures through the use of Finite Element Modeling (FEM). The benefit of this type of FEM modeling is its accuracy and the efficiency of its results.

EXPERIMENTAL SECTION

Accurate numerical modeling is essential for optimal fiber production and loss calculations. Comsol Multiphysics uses high-order vectorial elements along with an automatic and iterative grid refinement calculator for optimum error estimation. This process is a necessity when accurate silica-air interfaces need to be sampled during the meshing stage. The silica air interface puts a large burden on the computers memory because of fast far-field variations and the fact that the electric field's normal component will become discontinuous. Comsol Multiphysics uses a direct linear system solver (UMFPACK) for its modal analysis. This linear solver is commonly used to compute the equations involved in the FEM discretization of the Maxwell eigenproblem.

Random Hole Optical Fibers are very difficult to replicate in FEM software due to the various diameters and shapes of the air holes. In order to get an accurate structure when modeling using a FEM solver, Matlab must be used. Matlab has the capability of importing an SEM image and then converting it to grayscale. After import and conversion, Matlab can recognize the difference in pixels of the imported image. It is then a matter of exporting the file from Matlab into Comsol Multiphysics and changing the scale to match the original SEM data.

Figure 1 shown below is an SEM cross-section of the RHOF that has been created using the experimental fabrication process as described in[7]. The fiber has a 108.5µm outer diameter with an inner diameter at 64.6µm. The core region is approximately 20.3µm in diameter; the imported SEM image consists of thousands of air holes ranging in size from less than 100nm to above 1µm. The air holes non-uniform shape will also alter the birefringence. All models were examined at the free space wavelength λ_0=1550nm.

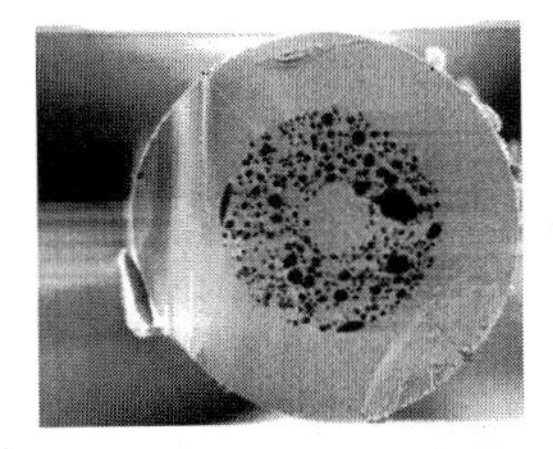

Figure 1. SEM cross section of Random Hole Optical Fiber.

FEM ANALYSIS

Once the SEM image in Figure 1 has successfully been imported into Comsol Multiphysics, the next step is to mesh the structure. The precision of the data directly relates to how refined the mesh can be. The only limitation of the mesh is the memory of the computer that will be solving the boundary equations. Figure 2 shows below a mesh consisting of 288892 elements run on a 4Gb Macintosh.

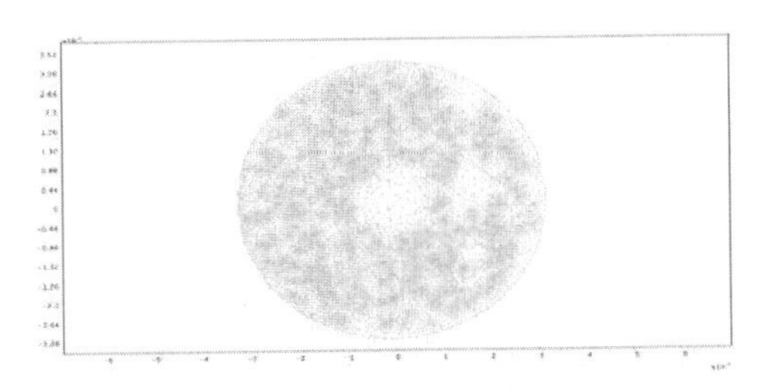

Figure 2. Mesh consisting of 288892 elements of imported SEM shown in Figure 1.

The outer diameter of the fiber is set as a cylindrical perfectly matched layer (PML), which utilizes a reflectionless outer layer. This is used in our research in order to absorb all outgoing waves. The PML is inherently a lossy material that is reflectionless for any frequency, thus absorbing all outgoing radiation. As shown in Figure 1, the actual outer boundary of the fiber is composed of a silica layer which is bulk fused to the random air holes. The PML layer gives leaky eigenmodes, which radiate toward the outer silica layer. No modes were expected in this region, and therefore, the outer layer has been omitted. This saved valuable memory space during the meshing and solving stages.

When the mesh is complete, the boundary conditions must be met. Continuity was set inside the cylindrical PML while a perfect electrical conductor was set to the outer edge. The air holes were then set to n=1.00 and the silica glass was set to n=1.45. Comsol is then solved using a direct linear equation solver (UMFPACK) near the effective mode index of the silica glass (1.449). Figure 3 shown below the resultant fundamental hybrid mode for the given SEM image in Figure 1. This is as expected for the power flow to be concentrated directly in the center of the core. The power flow time average z-component is at a maximum (in red) at 1.074×10^{17} W/m^2 and at a minimum (in blue) at 3.92×10^{-5} W/m^2.

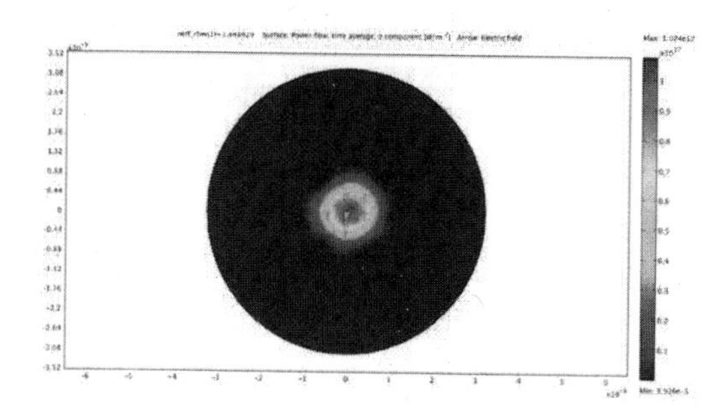

Figure 3. Transversal profile of the time-averaged power flow (W/m^2) in the z direction.

Figure 4, which is shown below, is a magnified picture of Figure 3; in order to make the electric field (green cones) more visible. The electric field in Figure 5, also shown below, gives the relative height (3D) of the power flow time average z-component. This relative height is proportional the optical intensity $|\mathbf{E}|^2$.

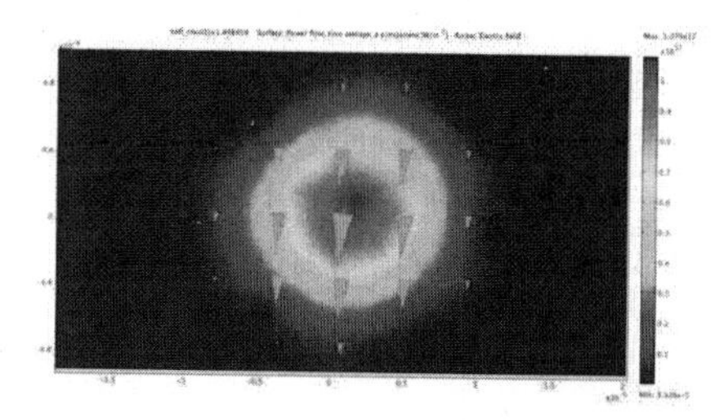

Figure 4. Fundamental mode of the fiber (magnified view).

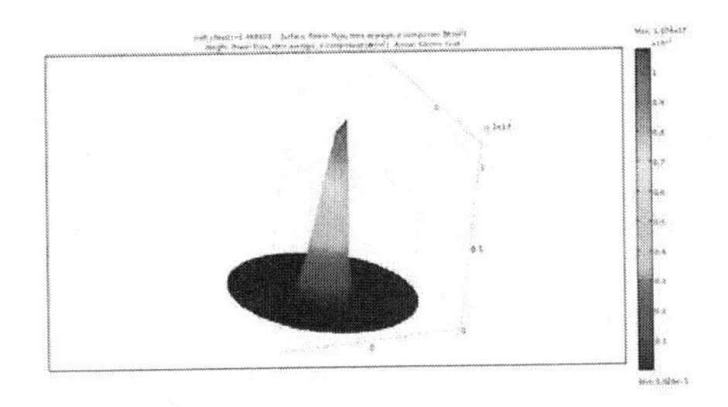

Figure 5. Power flow (W/m^2) 3D view.

Figure 6, as shown below, gives the various LP_{11} modes that also propagate in this multimodal fiber.

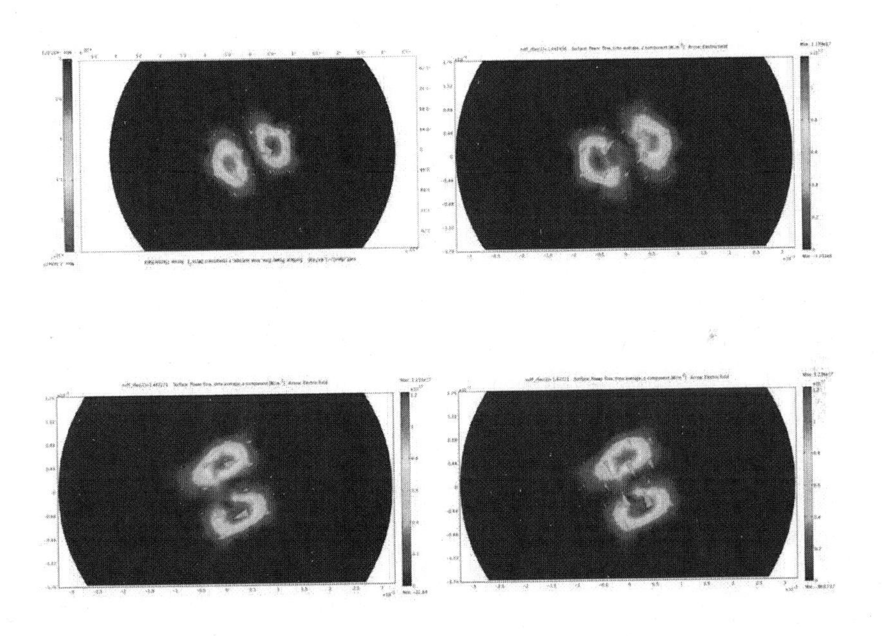

Figure 6. LP_{11} modes of the fiber near n=1.4495 (magnified view) clockwise from top left y-polarized even $\psi = A_0 J_1(ur)\cos\varphi \mathbf{a_y}$, x-polarized even $\psi = A_0 J_1(ur)\cos\varphi \mathbf{a_x}$, x-polarized odd $\psi = A_0 J_1(ur)\sin\varphi \mathbf{a_z}$, y-polarized odd $\psi = A_0 J_1(ur)\sin\varphi \mathbf{a_y}$.

CONCLUSIONS

This work presents the results of a modeling study of RHOF using a multiphysics modeling program. The importation process of an SEM cross-section simplifies the modeling process and provides a great degree of accuracy in the image translation using Comsol Multiphysics with Matlab. This process allows the user to easily modify the current air hole arrangement in order to further analyze any changes that can or will be made in the future.

REFERENCES

[1]G. Pickrell, W. Peng, and A. Wang, "Random-hole Fiber evanescent Wave Gas Sensing," Optics Letters, 29 [13] 1476-78, (2004)

[2]J. C. Knight, T. A. Birks, P. St. J. Russell, and D. M. Atk in, Opt. Lett. 21, 1547 (1996)

[3]T. A. Birks, J. C. Knight, and P. St. J. Russell, Opt. Lett. 22, 961 (1997).

[4]J. C. Knight, T. A. Birks, P. St. J. Russell, and J. P. de Sandro, J. Opt. Soc. Am. A 15, 748 (1998).

[5]T. M. Monro, P. J. Bennett, N. G. R. Broderick, and D. J. Richardson, Opt. Lett. 25, 206 (2000).

[6]B. J. Mangan, J. C. Knight, T. A. Birks, and P. St. J. Russell, in Digest of Conference on Lasers and Electro-Optics (CLEO), Optical Society of America, Washington, D.C., 559 – 560 (1999).

[7]G. R. Pickrell, D. Kominsky, R. H. Stolen, J. Kim, F. Elllis, A. Safaai-Jazi, and A. Wang, IEEE Photon. Technol. Lett. 16, 491 (2004).

INCREASED FUNCTIONALITY OF NOVEL NANO-POROUS FIBER OPTIC STRUCTURES THROUGH ELECTROLESS COPPER DEPOSITION AND QUANTUM DOT SOLUTIONS

Michael G. Wooddell, Dr. Gary Pickrell, Brian Scott
Materials Science and Engineering Department, Virginia Tech
Blacksburg, VA, USA

ABSTRACT

Two novel fiber optic structures have been developed that utilize phase separating boro-silicate glass to create a three-dimensionally interconnected pore structure. The first structure is similar to a "traditional" optical fiber where the porous glass has been used to replace the solid cladding material around a solid glass core. The second structure consists of an ordered array of nano-porous hollow tubes surrounding a hollow nano-porous core (termed hybrid fibers). The objective of this work is to increase the functionality of these optical fiber structures by integrating electronic and photonic materials onto the fiber surfaces and in the fiber pore structure. Conductive copper pathways were created on/in the fibers with a "traditional" configuration using electroless deposition. CdSe quantum dots were deposited from solution via capillary action into the hybrid optical fibers. Optical microscopy, SEM, EDS were used to characterize the fibers and their respective coating materials.

INTRODUCTION

The ever increasing needs of the human race for efficient, economical, and sustainable power generation have driven the demand for more accurate, versatile, and robust sensing devices and mechanisms. Fiber optic sensors are viable candidates to meet these demands due to their flexibility, range of possible applications, and sensitivity to a multitude of environmental parameters. They are capable of detecting and measuring nearly any environmental parameter including pressure, temperature, strain, and chemical composition. In addition they are not affected by electric fields and can maintain and operate at elevated temperatures, making them superior candidates for harsh environment sensing applications[1, 2].

The future of sensing technology for power generation applications will require devices that can detect numerous environmental parameters while retaining the ability to accurately transmit large amounts of information. Multifunctional fiber optic sensors and devices could provide a means to sense and transmit environmental conditions without the need for additional communication systems.

Traditional fiber optic sensors based on solid core/cladding configurations are often limited to single functions. Relatively new fiber optic structures such as photonic crystals, random-hole optical fibers, and porous clad fibers provide variations on the classic fiber optic structure that have the potential to increase the functionality of single strand fiber optic devices. Many of these new fiber optic structures have already demonstrated their ability to sense chemical, temperature, and pressure stimuli in industrial applications [3-11]. The application of photo and electoactive materials to fibers can further enhance their sensing capabilities as well as their ability to transmit data [12-15]. Combining microstructured and porous clad optical fibers with photo and electro active materials has the potential to introduce a new class of fiber optic sensors with multiple functions and the ability to monitor numerous environmental conditions while still retaining the ability to effectively transmit data. Thus it is the objective of this segment of the research has been to confirm the feasibility of depositing various materials on nano-porous clad glass fibers for use in sensing and electronics applications. The basic deposition methods and characterization can be utilized as building blocks to create electronic device components and active coating elements within and on the surface of the fibers. More specifically the electroless deposition of copper has been demonstrated as a means to create multifunctional fiber optic sensors.

Additionally, the manufacturing of stochastic ordered hole fibers, or fibers that contain both a patterned array of air holes running longitudinally along the fiber and random porosity in the bulk glass material has been demonstrated. Semiconductor quantum dots have been deposited in the stochastic ordered hole fibers to highlight the viability of using the coated fibers in sensing applications.

FIBER FABRICATION

The fibers used in this study were produced in house using standard fiber drawing techniques and chemical treatment regimes for creating porous glass. The fibers were created using a 4 stage process. The stages were preform manufacture, fiber drawing, heat treatment, and chemical treatment

Preforms for the porous clad fibers were constructed by collapsing a tube of phase separable "green" vycor brand glass around a non phase separable glass core. In this case, the term "green" is used to define the glass state before phase separation. Fibers were then drawn from the perform using standard fiber pulling techniques which employ a glass lathe and oxygen/hydrogen torch. The process used to create ordered hole fiber performs and fibers differed greatly from that of the porous clad fibers.

Figure 1 presents a schematic flow chart of the process used to create the porous ordered hole fibers. The process utilizes what is termed the stack and draw method to produce fibers. First, tubes of "green" vycor glass approximately 7mm in diameter were drawn down to produce smaller diameter tubes of approximately 300-1500 μm using a rotating glass lathe and a jeweler's torch with oxygen hydrogen flame. Tubes of similar size, (ranges of 50μm) were separated and grouped together. The tubes of reduced diameter were then stacked into another tube of "green" vycor glass whose inner diameter was approximately three times the outer diameter of one of the reduced diameter tubes. This would produce a structure with six hollow tubes surrounding a seventh central tube as shown in Figure 1. Next, a rod of borosilicate glass was fused to the tip of the preform. Finally, a small area behind the fused region was heated with the jeweler's torch with a small low heat flame as both preform and rod rotated in the glass lathe. The purpose of the small low heat flame was to provide even heating across the entire heated zone and also prevent the preform from collapsing on itself. When the preform was evenly heated, the flame was removed, the lathe rotation was stopped, and the borosilicate rod was pulled away from the fiber preform.

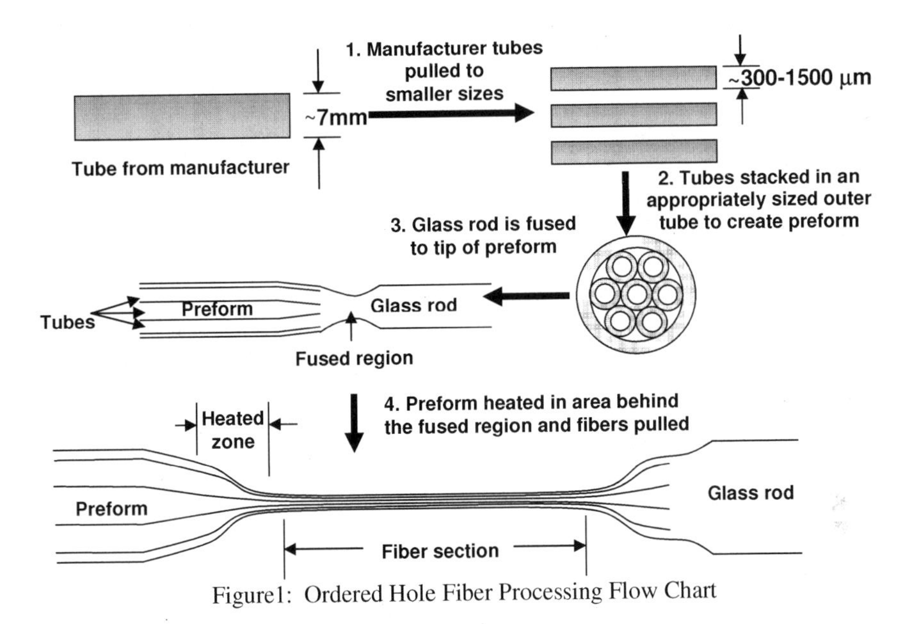

Figure1: Ordered Hole Fiber Processing Flow Chart

The drawing process for the porous clad fibers yielded fibers with lengths ranging from 3-10 ft. The 3-10 ft fibers were then sectioned into 12 in. pieces for the subsequent processing steps. The drawing process for the ordered hole fibers yielded fibers with lengths ranging from 6 in. to 2ft. These fibers were then sectioned into small pieces approximately 1 in. in length. All fibers were heat treated between 565 and 600 °C for a period of 24 hours to induce the spinodal phase separation, and then were allowed to furnace cool. Upon cooling, the fibers were placed in a 5 % ammonium biflouride solution for five minutes to initiate the dissolution process. This step helped to remove the surface layer of material and provide access to the borate rich regions of the fiber. From here the fibers were placed in a 3 N nitric acid solution at a temperature of 85-90 °C for 22-24 hours. Upon completion of the leaching process, the fibers were rinsed in deionized water for 24 hours to remove residual traces of acid solution. Finally, the fibers were dried at 100 °C for 12 hours to remove residual water from the pore structure.

ELECTROLESS DEPOSITION OF POROUS CLAD FIBERS[16]

Good adherence of the copper coating requires a clean substrate free of organic residue that might be present due to handling. Thus the fibers were first cleaned in a series of steps. To start, the fibers were ultrasonically cleaned in an ethyl alcohol bath for 15 min, followed by a 15 minute rinse in DI water and subsequent drying for 20 min at 100 °C. This rinsing and drying process was performed after each processing step to remove residual solution from previous baths. This allows for new solutions to easily access the pore structure and prevents bath contamination, as this is a crucial component to preventing the crash of the activating/plating baths. Following the ultrasonic cleaning, the fibers were cleaned in an acidic solution containing 75 ml DI water and 25 ml HCl

36.5-38% for 15 minutes at 30-35 °C. The cleaning in acidic solution aimed to remove any residual organic matter that might be present on the fibers from handling.

Fibers were sensitized using a solution of $SnCl_2$. The bath consisted of 120 ml DI water, 3.0g $SnCl_2$, and 5 ml HCl. Fibers were treated in the sensitizing solution for a period of 15 minutes at room temperature. Activation of the fibers was performed in a bath consisting of 125 ml DI water, 0.03g $PdCl_2$, and approximately 0.063 ml HCl for 25 minutes at 45 °C.

The plating process was completed in a series of steps, the first of which was the production of the plating solution. Plating solution A consisted of 30 ml DI water and 5.0g of NaOH. The purpose of this solution was to simply adjust the pH of plating solution B, which consisted of 80 ml DI water, 1.25 g $CuSO_4 \cdot 5H_2O$, and 7.5 g of Ethylenediaminetetraacetic acid (EDTA). Solution B was brought to a temperature of 40 °C and stirred at 400 rpm while half solution A was added. This gave the solution a deep blue color and a pH of approximately 9. The remainder of solution A was added to bring the pH up to approximately 12.5. At this point, the solution was not yet ready for plating as it lacked the reducing agent, formaldehyde. When ready to plate, the mixture of plating solutions A and B was brought to a temperature of 45 °C and 2.5 or 5.0 ml of HCOH 37 wt% in H_2O was added. Fibers were placed in the plating solution for periods of 20 and 25 minutes.

Experimental Parameters

Experimental parameters in the Cu plating experiments were plating time, HCOH concentration, and post heat treatment (HT) time of fibers. The plating time and HCOH parameters were adjusted to investigate their effects on coating quality and electrical properties of the fibers, while different heat treatment times were investigated to determine the effects on the depth of penetration of copper into the pore structure. Table 1 presents how the parameters were varied. Three samples of each experimental set were produced in order to determine if the results were repeatable.

Table 1 Electroless Plating Experimental Parameters

	Plating Time, min	
Heat Treatment Time, min	20	25
15	2.3 and 4.4% HCOH	2.3 and 4.4% HCOH
25	2.3 and 4.4% HCOH	2.3 and 4.4% HCOH

Electrical Characterization of Copper Plated Fibers

The Cu plated fibers were fragile after the processing described in section 2 above. In order to test them electrically they had to be placed on a rigid substrate and held down with conductive silver paint. The conductive silver paint would act as the measurement points between the fibers. The setup is shown schematically in Figure 2.

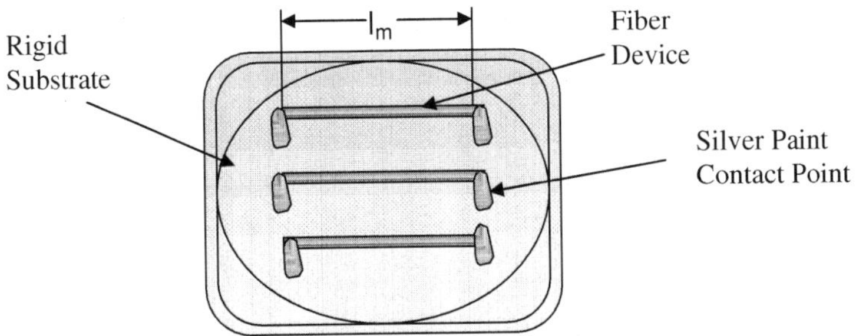

Figure 2: Copper Plated Fibers Test Apparatus

Resistance per unit length measurements were made by placing probes from a digital multimeter onto the silver paint connected to the fibers. The measurement distance, l_m, was the distance between points where the silver paint contacted and was chosen to be the effective length of the device.

STOCHASITC ORDERED HOLE FIBERS

The term stochastic ordered hole fiber (SOHF) refers to the combination of MOFs and the porous clad fibers discussed above. The combination of these different types of fiber optic structures provides a truly three dimensional microstructure that can utilize the advantages of each individual structure. Figure 3 presents a schematic representation of one such stochastic ordered hole fiber geometry (obvisouly many such structures are possible). Fibers such as these allow for chemical and gas transport both axially and longitudinally through the fiber. These types of fibers could further reduce the response time of chemical and gas sensors due to shorter diffusion distances and increased transport area, in addition to stronger interaction with the propagating modes. Other than the work completed at Virginia Tech[17-19], there is no other data available in the literature detailing attempts to create such fibers. Thus, it is believed that this is a novel fiber structure developed at Virginia Tech.

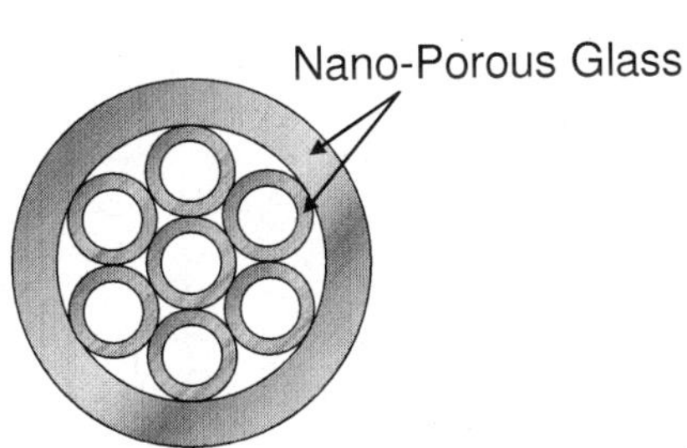

Figure 3: SOHF. Contains elements of both MOF's and Porous Clad Optical Fibers

Stochastic ordered hole fibers which contain characteristics of MOFs and porous clad fibers were produced using the stack and draw method, then heat treated and leached similar to those fibers containing only a porous cladding as described above. Figure 4 presents a section of a SOHF magnified to 100x.

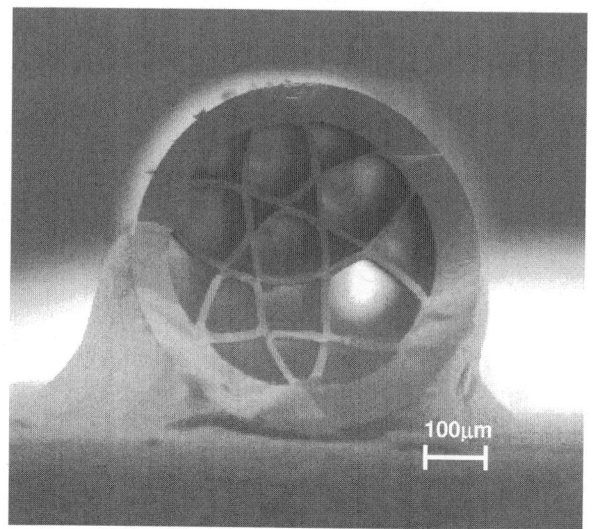

Figure 4: SOHF Fiber at 100x

The micrographs show that, although the six ring pattern around the central tube is not perfect, the overall pattern was retained from the preform throughout the drawing operation. A higher magnification image of this fibers is shown in Figure 5. This image indicates these fibers contain the nano-porous structure seen in the porous clad fibers shown previously.

Figure 5: SOHF at 50 kx Mag

ELECTROLESS PLATING RESULTS

The objective of the electroless plating of copper experiments was aimed at creating a uniform coating on the surface of the porous clad fibers, depositing copper within the pore structure, and determining variable processing conditions that could offer a range of electrical properties. SEM analysis of the fibers provides evidence that the fibers have been coated with a uniform layer of Cu, and that the Cu has plated in the pore structure.

Figure 6 presents an SEM micrograph of a Cu plated fiber with no HT, 2.3% HCOH, and 25 minute plating time. From Figure 6 it is possible to see that the Cu coating has a nearly uniform thickness of approximately 20 μm on the fiber surface.

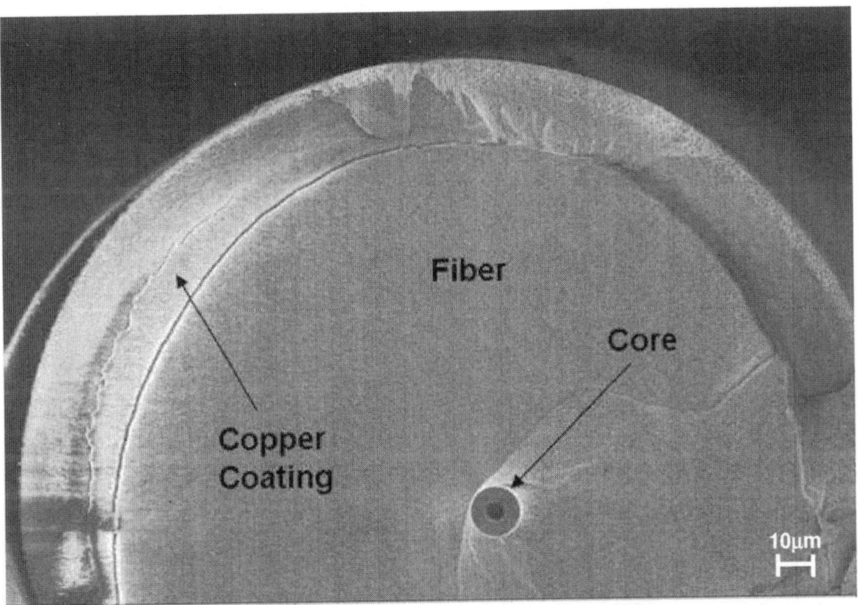

Figure 6: Cleaved Cu Coated Fiber Face 600x

Also visible in Figure 6 is what appears to be a hole at the center of the core region. This hole is present due to incomplete collapse of the borosilicate tube used to create the core. The hole in the core is not detrimental to the fiber structure; instead it shows that it possible to create fibers with both solid cores, as well as cores with a tube running longitudinally through its central axis.

The interface region between what is termed the bulk copper plating, or the copper on the surface of the fiber, and the copper coated pore structure is of interest because it will have an influence on the adhesive properties of the coating. Figure 7 presents an SEM micrograph of the interface between the bulk Cu plating and the Cu plated pore structure.

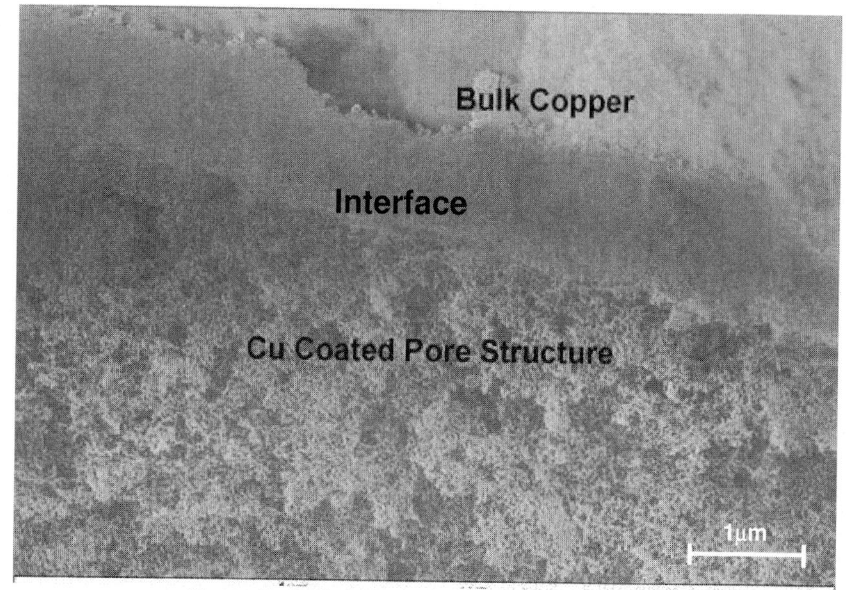

Figure 7: Bulk Cu/Pore Structure Interface 30kx

In Figure 7 it is possible to see that there is a relatively sharp interface between the bulk copper coating and the copper coated pore structure.

Despite the uniform coating depicted in Figure 6, some Cu coatings displayed poor adherence to the glass fiber substrate. The poor adherence to some fibers could have resulted from contaminated surfaces as well as NaOH attack of the silica network during the plating process. Additionally, the sharp interface between the copper coated pore structure and the bulk copper coating might also lead to scaling of the coatings.

Plating Depth v. Heat Treatment Time

Theoretically the electroless plating of Cu should occur where ever the activated surface comes in contact with the plating solution. However, EDS mapping shows that the Cu penetrates only a short distance into the pore structure of porous cladding. The likely culprit preventing the complete plating of the fibers is transport of Cu^{2+} ions and HCOH molecules into the porous structure and as well as the transport of $HCOO^-$ and H_2 molecules out of the porous structure. This could be used to make gradient structures of copper which vary in concentration from the surface to the core of the fiber. As the plating solution begins to enter the porous structure, the regions closer to the fiber surface begin to plate while solution is still being transported to the center of the fiber. Plating of the pore structure closer to the surface causes the pores to become clogged thus limiting further transport of reacting species to the inner pore surfaces. Increasing the pore size in the cladding region may help to increase transport of plating solution into the pore structure and also increase the time between total pore blockage, thus increasing the penetration depth of the Cu coating. Figure 8 presents the plot of Cu x-ray counts using EDS v. radial distance from the pore surface for fibers with no heat treatment and heat treatments of 15 and 25 minutes at 750 °C (note: fibers were plated for 25 minutes in 2.3 % HCOH solution).

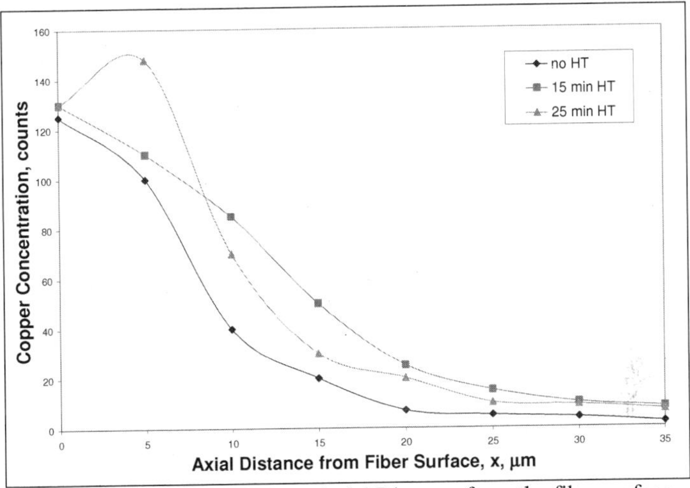

Figure 8: Cu X-ray Counts v. Axial Distance from the fiber surface

From the plot it is possible to see that the fiber with no heat treatment has a dramatic decrease in Cu counts a shorter distance from the surface compared to the heat treated fibers. At 15 μm the un-heat treated fiber shows Cu x-ray counts of 20, while fibers heat treated for 15 and 25 minutes show counts of 30 and 50 respectively. At approximately 35 μm, the Cu counts for the heat treated fibers level off at 10. The un-heat treated fiber reaches this value at approximately 20 μm. It can be concluded that the heat treated fibers were plated approximated 15 μm further into the pore structure.

QUANTUM DOT DOPED STOCHASTIC ORDERED HOLE FIBERS

Quantum dots (QDs) show intense fluorescence when exposed to various types of radiation. The CdSe quantum dots with a shell of ZnS are highly fluorescent when exposed to UV radiation. Figure 9 shows a low magnification optical image of stochastic ordered hole fibers that have absorbed quantum dots into their tube structure. The size of the quantum dots, between 2 and 10 nm, enables them to enter the pore structure of the fibers. Each color of the QDs in Figure 9 represents a different QD size, the yellow being 2.4 nm, the orange being 4.0nm, and the red being 5.2 nm The smaller the quantum dots, the smaller the wavelength of the radiation that they emit. Therefore, the smaller QDs emit light more towards the blue side of the spectrum and the larger QDs emit more towards the red side of the spectrum. This is evident when looking at Figure 10, which shows the same three fibers in Figure 9 illuminated under long wavelength UV radiation (364nm wavelength). The colors emitted are green, orange-yellow, and red.

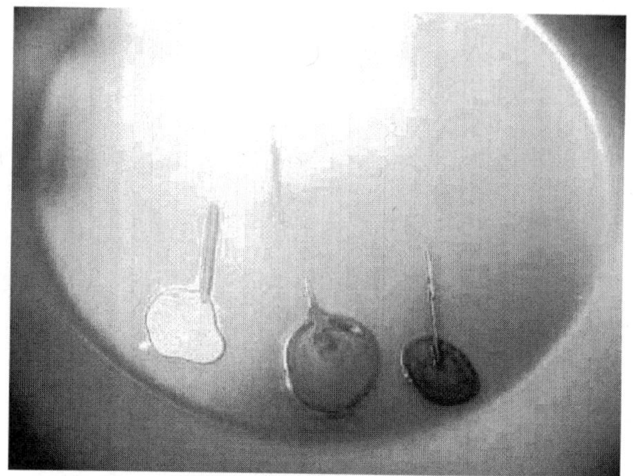

Figure 9: SOHF with QDs

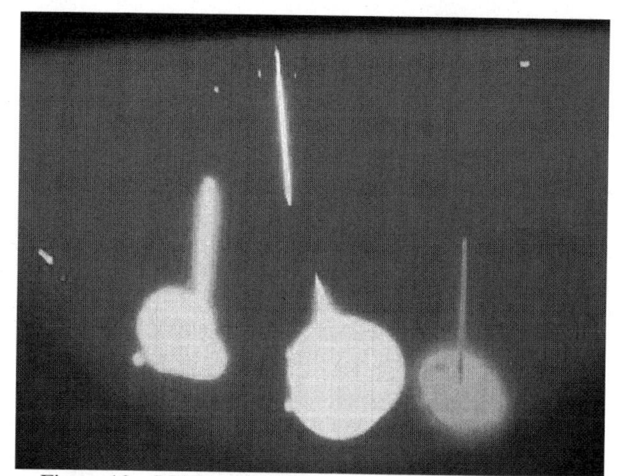

Figure 10: SOHF with QDs under 364 nm illumination

ACKNOWLEDGEMENTS

The financial support of the US Department of Energy's National Energy Technology Laboratory (NETL) under award number DEFC2605NT42441 is gratefully acknowledged.

REFERENCES:

1. Udd, E., ed. *Fiber Optic Sensors: An Introduction for Engineers and Scientists.* 1991, John Wiley and Sons, Inc.: New York
2. K.T.V Grattan, B.T.M., ed. *Optical Fiber Sensor Technology.* 1995 Chapman and Hall: London.
3. Didier Felbacq, S.G., Boris Kuhlmey, Andre Nicolet, Gilles Revnersez, Frederic Zolla, *Foundations of Photonic Cyrstal Fibers.* 2005, London: Imperial College Press.
4. Zhu, Y. *Long-period gratings in photonic crystal fiber: Fabrication, characterization, and potential applications for gas sensing.* 2005. Boston, MA, United States: International Society for Optical Engineering, Bellingham WA, WA 98227-0010, United States.
5. Petrovich, M.N., et al. *Microstructured fibres for sensing applications.* 2005. Boston, MA, United States: International Society for Optical Engineering, Bellingham WA, WA 98227-0010, United States.
6. Minkovich, V.P., et al., *Microstructured optical fiber coated with thin films for gas and chemical sensing.* Optics Express, 2006. **14**(18): p. 8413-8418.
7. Ho, H.L., et al., *Optimizing microstructured optical fibers for evanescent wave gas sensing.* Sensors and Actuators, B: Chemical, 2007. **122**(1): p. 289-294.
8. Jensen, J.B., et al., *Photonic crystal fiber-based evanescent-wave sensor for detection of biomolecules in aqueous solutions.* Optics and Photonics News, 2004. **15**(12): p. 16.
9. Pickrell, G., W. Peng, and B. Alfeeli. *Fiber optic chemical sensing.* 2005. Boston, MA, United States: International Society for Optical Engineering, Bellingham WA, WA 98227-0010, United States.
10. Pickrell, G., et al., *Microstructural Analysis of Random Hole Optical Fibers.* IEEE Photonics Technology Letters, 2004. **16**(2): p. 491-493.
11. Pickrell, G., W. Peng, and A. Wang, *Random-hole optical fiber evanescent-wave gas sensing.* Optics Letters, 2004. **29**(13): p. 1476-1478.
12. Fox, G.R. and D. Damjanovic, *Electrical characterization of sputter-deposited ZnO coatings on optical fibers.* Sensors and Actuators, A: Physical, 1997. **63**(2): p. 153-160.
13. Jianming, Y., et al. *Fiber optic chemical sensors using a modified conducting polymer cladding.* 2001. Boston, MA, USA: SPIE-Int. Soc. Opt. Eng.
14. Yuan, J. and M.A. El-Sherif, *Fiber-optic chemical sensor using polyaniline as modified cladding material.* IEEE Sensors Journal, 2003. **3**(1): p. 5-12.
15. Fox, G.R., et al., *Piezoelectric coatings for active optical fiber devices.* Ferroelectrics, 1997. **201**(1-4): p. 13-22.
16. Poquette, B., *Damping Behavior in Ferroelectric Reinforced Metal Matrix Composites*, in *Materials Science and Engineering.* 2005, Virginia Polytechnic Institute and State University: Blacksburg. p. 45.
17. Brian L. Scott, C.M., Kristie Cooper, Anbo Wang, T. Ooi. *Novel Chemical and Biological Fiber Optic Sensor.* 2007: SPIE- The International Society for Optical Engineers.

18. Gary Pickrell, B.S., Cheng Ma, Kristie Cooper, Anbo Wang, *Stochastic holey optical fibers for gas sensing applications*, in *Optics East: Sensors for Harsh Environments III*. 2007, SPIE- The International Society for Optical Engineering: Boston.
19. Wooddell, M., G. Pickrell, and T. Ooi. *Development of Stochasitc Optical Fibers for Structural Health Monitering Applications*. in *Structural Health Monitering* 2007.

THERMOPOWER MEASUREMENTS IN 1-D SEMICONDUCTOR SYSTEMS

Sezhian Annamalai, Jugdersuren Battogtokh, Rudra Bhatta, Ian L. Pegg and Biprodas Dutta

Vitreous State Laboratory
The Catholic University of America
Washington, DC 20064.

ABSTRACT

Measurement of thermopower (Seebeck coefficient) in one dimensional systems is substantially more complicated than similar measurements in bulk materials. In micro/nano dimensions, the high resistances of semiconductor/metal interfaces, contact resistance, have to be negotiated to perform meaningful experiments. Thermopower of a material is the characteristic voltage generated when a temperature differential is maintained across the length of that material. During the thermopower measurement of bulk materials, a voltmeter is used t measure the voltage generated, whose internal resistance is substantially greater (~ 10^6 times) than that of the sample. For prospective thermoelectric materials such as PbTe, the resistance of a 1 μm dia wire could be as high as 10^6 Ohm, ncccssitating a voltmeter with an internal resistance of ~ 10^{12} Ohm. Moreover, the unknown contact resistances at the metal electrode - semiconductor micro/nano wire interfaces result in conflicting thermopower data reported by different laboratories. Several issues arising out of thermopower measurements in 1-*d* semiconductor systems and measures to counter these have been discussed.

INTRODUCTION

When a temperature difference (ΔT) is maintained across the length of any material, charge carriers are transported from the hot to the cold side, creating an open circuit voltage (ΔV) or potential difference. Such a potential difference is proportional to the temperature difference (Eqn. 1) and the proportionality constant is called Seebeck coefficient (S) or thermopower (Eqn. 2). If a load is connected in the external circuit, and if the current flows from the cold to the hot end of the sample, then S is considered positive (as a convention).

$$\Delta V \, \alpha \, \Delta T \qquad (1)$$

$$S = \frac{\Delta V}{\Delta T} \qquad (2)$$

In metallic or highly conducting bulk materials, thermopower measurement follows a relatively straight forward procedure. The two ends of the sample (preferably, in the form of a wire) are brought intimately in contact with a heat source and a heat sink to serve as the hot and cold ends, respectively. The contact with the heat source and sink is facilitated by some form of electrically and thermally conductive paste such as colloidal silver. Vapor deposition of silver, carbon or sputtering to attach the two ends of the sample with the source/sink is complicated and also unnecessary for bulk specimens. The next step in the measurement procedure is the application of a temperature difference between the sink and source followed by measuring the potential difference generated across the two ends of the sample. To determine the Seebeck coefficient using Eqn. (2), the temperature difference between the hot and cold ends must be known precisely. A variety of methods can be employed to determine ΔT, the most prevalent being thermocouples embedded in the source and the sink.

Thermopower measurement procedures in micro/nano semiconductor wires (one-dimensional or 1-*d* systems) are, in many ways, different from those adopted during measurements in three-dimensional (bulk or 3-*d*) metallic systems. The complexities in the former arise from a variety of sources such as: i) impurity or an oxide phase on the semiconductor surface, ii) work function mismatch between semiconductor and metal, etc. As a result of all such problems, a variable contact resistance is introduced in the system. Another issue, typical of electrical and thermal measurements of 1-*d* systems, is the state of bonding of the electrode material with the semiconductor. Normally, this is taken care of by subjecting the sample to an annealing step at a suitable highly temperature to form a compact bond between the metal and the semiconductor. In general, problems arising out of contact resistance are more acute in the case of 1*d* micro/nano wires (MNWs) compared to bulk materials as there are many low resistance pathways possible for charge carriers in the latter.

In general, thermopower in most materials is a function of temperature. However, thermopower of semiconductors exhibit far more sensitivity towards temperature than metals. Hence, for semiconductors, the temperature difference or the imposed ΔT should not be large and is normally kept in the range 1 – 3 K. Considering that the thermopower of most materials are relatively low (a few hundred microvolts per Kelvin), very reliable voltmeters must be resorted to for measurements of such low voltages. The problem of low voltage measurement is compounded by the fact that as the diameters of MNWs approach the submicron (< 1000 nm) range, the sample resistance increase considerably. Such high resistance often leads to very weak signals, which add immensely to the complexity of measurement procedure, necessitating the use of triaxial cables, shielding the entire experimental setup to avoid spurious extraneous currents and or voltages. Thermopower measurements of MNWs can basically be thought of measuring the voltage generated by a battery which has an extremely high internal resistance.

In order for a voltmeter to measure the voltage generated by a high resistance material such as a semiconductor MNW, it is mandatory for the voltmeter to have its internal resistance several orders of magnitude greater (preferably 10^6 times) than that of the sample. In such circumstances, the current flowing through the voltmeter will be negligible, compared to that flowing through the sample since the two are connected in parallel. In the case of semiconductor 1*d* MNWs, the sample resistance could be of the order of 1 GΩ at room temperature. Such resistances necessitate the internal resistance of the voltmeter to be at least 10^{15} Ω. Even though there are some voltmeters available commercially with such high internal resistances, the instrumental errors in such instruments are of the order of 350 μVolt (or $\pm 350 \times 10^{-6}$ Volts). As the thermopower (EMF) generated by most prospective semiconductor MNWs (with $\Delta T \sim 3$ K) is of the order of the instrumental error, reliable thermopower determination in by direct voltage measurements is difficult.

In order to circumvent the problems associated with voltage measurements in 1*d* MNWs, an alternative measurement technique is presented which is based on measuring the current generated by the MNWs when a ΔT is imposed. If the resistance of the sample is measured accurately, the Seebeck coefficient or thermopower generated across the length of the sample can be determined indirectly by multiplying the current with the resistance. This method seems useful since electrometers with very low current measurement capabilities are widely available. In fact Keithley model 6430 source-meter has been employed to determine thermopower in MNWs as described in a later section.

Another issue which complicates thermopower measurements in 1*d* MNWs is that of Joule heating when a current is generated during the experiment. The current, I, initiates Joule heating which is proportional to I^2R, where R is the resistance of the MNW. While this is not a concern for metallic or highly conducting material, it poses a serious problem for semi conductors who are also poor conductors of heat. The temperature of the MNWs can increase rapidly because of their inability to dissipate the heat away, resulting in melting.

A comprehensive technique for measuring thermopower in semiconductor micro/nanowires is presented which mitigates the problems discussed above.

EXPERIMENTAL PROCEDURE, RESULTS AND DISCUSSION

A standard thermopower measuring set up is shown in Figures 1A and 1B. It consists of a thin alumina substrate (6) (<0.5mm thickness) on which three massive (compared to the size of the sample wire) copper pads (2 and 3) are attached which act as heat source and sink. A micro heater (1) is located closer to the heat source and is used to generate a temperature difference across the sample as well as the reference wire. The reference wire is a NIST traceable constantan wire (1/100 inch in diameter) of known S, which is used to measure ΔT. Even though two separate Cu pads are designated as the heat sources for the sample and the reference, the Cu pad acting as the heat sink is common.

A copper block (7) is positioned below the alumina substrate (Figs. 1A-B), which is placed on a ceramic heater (8). Such a configuration allows the heat to be conducted uniformly to the sample and reference material. This arrangement also ensures that the sample and the reference are maintained at the same temperature. Another copper block (9) is attached to the alumina substrate beneath the microheater which helps in maintaining uniform temperature around the heater. In this set up, the sample (5) is placed between 2 and 3 while the reference wire (4) is placed across 2 and 3 as shown in Fig 1A. With the help of the ceramic heater the temperature of the entire setup can be raised to the desired temperature up to a maximum of 750 K. Once the system is stabilized at the desired temperature, the background voltage of the system (initial voltage) is measured across the reference wire (V_{IR}). Voltage measurement across a constantan wire is relatively simple because its resistance is extremely low and an ordinary voltmeter (internal resistance ~ 1 GΩ) serves the purpose adequately. Even though the source of the initial voltage is not precisely known, it arises out of the junctions formed from different metals and conductive glues used to attach electrodes on the sample and also from the electronics in the electrometers/multimeters and power sources used in the measurement system.

When the microheater (~ 100 Ω resistance heater) is activated by passing a current through it by supplying about 100 – 200 mW power, the source and sink receive the same heat flux from it because they are located symmetrically vis-à-vis the position of the heater. Moreover, the sink is placed further away from the source to ensure that the sources are maintained at a higher temperature relative to that of the sink. If sufficient time is allowed (e.g. 5 minutes), equilibrium is reached and a stable temperature difference between the sources and sink is achieved. Hence, by activating the microheater, a ΔT can be imposed between the ends of the sample as well as the reference. Imposition of a temperature difference generates a thermo-EMF (V_{FR}) or the final reference voltage. As the reference material is constantan and its thermopower, as a function of temperature (Table 1), is made available in the public domain by NIST, measurement of the constantan thermopower (voltage) allows determination of ΔT as follows:

$$\Delta T = \Delta V_R / S_R \qquad (3)$$

where $\Delta V_R = (V_{FR} - V_{IR})$ and ΔT is the temperature difference imposed by the micro heater and S_R is the NIST calibrated Seebeck coefficient of the reference material at the test temperature.

As described earlier, the temperature difference across the sample as well as the reference wire are the same due to the geometry adopted during placement of the heat sources and sink where the hot ends of the sample and the reference wire are located at the same distance from the heater while the sink is shared together by the sample and the reference wire. This symmetrical positioning of the sample and reference wires results in the generation of the same temperature difference across the sample and the reference wire. The measurement description provided in this article is applicable for a wide range of semiconductor compositions such as CdTe, PbTe, GsAs, GaSe, etc. Hence, the discussion will not be restricted to any particular semiconductor composition. Instead, the discussion will refer to the generic semiconductor micro/nanowire (MNW).

Thermopower Measurement of Micro/Nanowires

The 1-*d* MNWs used in the present investigation are composite glass-clad semiconductor wires, which are co-drawn in a optical fiber draw tower, using a proprietary method. The outer dimension of these fibers are typically 225 μm while the size of the inner semiconductor core can be varied between 100 μm to 100 nm. The advantage of these materials is that the glass cladding can be etched away by using hydrofluoric acid (HF) either to expose the semiconductor or to fully dissolve the glass cladding to get fully bare MNWs as depicted in Figures 2 – 3. Figure 2 exhibits a 4 μm MNW, which was etched in HF for 5 minutes to etch the ends of the composite fiber and expose the MNW tips. Fully bare semiconductor wires are shown in Figs. 3, which were prepared by longer etching times. HF does not seem to affect the semiconductor core significantly.

A typical measurement setup using a 6 μm bare semiconductor wire is shown in Fig. 4. The constantan wire (1/100 in thick) and the semiconductor MNW (faintly visible) are exhibited. The silver paste used for attaching the constantan and semiconductor MNW at the hot and cold ends can be observed in this figure. The utility of the constantan wire in the measurement procedure has been described earlier. The procedure for determining the thermopower of the semiconductor MNW is described in the following.

A Keithley 6430 source meter which has an internal resistance of 10^{-19} Ω as an ammeter was used. Triaxial cables were used to avoid signal loss and interference by spurious signals between sample chamber and measuring instrument. The middle shielding of the triaxial cable served as the guarding electrode by maintaining a voltage very similar to the core (conductor) of the triaxial cable. The outer shielding was connected to a ground connection dedicated for the measurements to avoid ground loops.

As described earlier, direct measurement of thermo-EMF will not be performed. The current generated by the imposition of a ΔT will be measured along with the resistance of the measurement circuit. As described during the thermopower measurement of the constantan wire, an initial current flows in the measurement circuit even without the imposition of a ΔT. After the initial current (I_{IS}) is measured, ΔT is imposed as described earlier. After allowing the system to equilibrate for ~ 5 min, the stable final current (I_{FS}) is measured. It is assumed that the increase in the current flowing in the circuit is caused exclusively by the difference in temperature imposed on the hot and cold ends of the sample. In other words, the excess current (I_{FS} - I_{IS}) is set up by the thermopower of the sample. It is also assumed that the resistance of the sample remains unaltered before and after ΔT is imposed. This assumption appears to be reasonable since the temperature difference is kept at a bare minimum.

The resistance of the sample and the entire measurement circuit is determined by Keithley 6430 by imposing different currents until a stable value of resistance is obtained over a large range of current values. Once the resistance (R_S) at a certain temperature is determined, the thermopower is determined by:

$$S = (I_{FS} - I_{IS}) * R_S / \Delta T \qquad (4).$$

Often times, the initial current in the circuit is too small to be accurately measured even with an ammeter capable of measuring currents in the 10^{-15} ampere range. In such cases, the final currents were measured at two ΔT and unambiguous determination of thermopower was possible even without measuring the initial current in the circuit as described in the following:

$$S_k = \Delta V_k / \Delta T_k \qquad (5)$$

$$S_k = S_j \qquad (6)$$

where the subscripts k and j indicate different ΔT's. Eqn. (6) states that a given material will have the same Seebeck coefficient at a given temperature irrespective of the temperature difference used for measuring it.

It follows from Eqns. 5 and 6:

$$\Delta V_k = V_{ki} - V_{kf} = \Delta V_j (\Delta T_k / \Delta T_j) \qquad (7)$$

where the second set of subscripts i and f, refer to the measurements done before and after the application of the temperature difference. Since the initial voltage or current measurement is the same ($V_{ki} = V_{ji}$) for both ΔTs, as this is the background and should be independent of ΔT, Eqn. (7) can be rewritten as

$$(V_{ki} - V_{kf}) = (V_{ki} - V_{jf})(\Delta T_k / \Delta T_j) \qquad (8)$$

Rearranging Eqn. (8):

$$V_{ki}(1- (\Delta T_k / \Delta T_j)) = (V_{kf} - V_{jf}(\Delta T_k / \Delta T_j)) \qquad (9)$$

Finally the voltages can be replaced by the currents using Ohm's law as shown below:

$$V_{ki} = R*(I_{kf} - I_{jf}(\Delta T_k / \Delta T_j))/(1- (\Delta T_k / \Delta T_j)) \qquad (10)$$

$$V_{kf} = R * I_{kf} \qquad (11)$$

Using V_{ki} and V_{kf} and the temperature difference calculated for the reference material the thermopower of the sample can be calculated using Eqn. (5).

Joule Heating of MNWs

To measure the resistance of an MNW, a current has to be passed through it. Joule heating is a natural consequence when a current is passed through it. If the heat generated by Joule heating is dissipated to the ambience inside the evacuated sample chamber, the measurement is unaffected. However, if the heat is not dissipated, the temperature of the MNW rises and interferes with the measurement. The amount of heat generated is given by I^2R, where I is the current passed and R is the resistance of the sample. A stable value of the resistance may be obtained but the measurement temperature becomes uncertain. This problem is severe in the case of semi-conductor MNWs which have much higher resistances compared to a metal. As depicted in Figs. 5A and 5B, a semiconductor MNW may melt and even evaporate as a result of even a moderate current passing through it. The sample in Figs. 5 A and B is CdTe MNW with a diameter of 10 μm. The sample burnt after a 0.16 mA current was passed through the wire. For thinner samples, a much lower current may melt and the sample may even evaporate without leaving any trace in the sample holder. In the case of a glass-clad MNW, the results of Joule heating could be more problematic since the heat generated remains trapped in the glass, and it would also be difficult to observe the failure because of the glass cover.

One way of determining the onset of Joule heating is as follows. In the Ohmic region (voltage proportional to current) the resistance of the wire remains constant as the current is increased gradually. After a certain current, the resistance will be observed to decrease (Fig. 7) monotonically with increasing current. The onset of Joule heating is considered to be the reason for this decrease in resistance. Measurement of sample resistance must be performed by passing currents which are at least one order of magnitude lower than the current where Joule heating ensues. The maximum current that can be used for determination of resistance depends on the volume of the sample holder, the material it is made of, the vacuum level maintained during the experiment, the emissivity of the MNW, the temperature of measurement and several other factors. For most semiconductor MNWs, a rule of thumb successfully devised on the basis of repeated experiments is that the current density should not exceed 10 Amp/cm^2.

CONCLUSIONS

A novel method for the determination of thermopower in 1-*d* semiconductor micro/nanowire systems has been presented. Major problems encountered are measurement of feeble signals and Joule heating. Effective measures have been undertaken to mitigate both problems to ensure reproducible thermopower measurements. Effective measures to avoid Joule heating of the samples have also been discussed.

ACKNOWLEDGEMENTS

We gratefully acknowledge the contributions of Dr. Robert Mohr, Ms. Wei Zhou, Dr. Igor Vidensky, and Dr. Thomas Barnard in this investigation.

Table 1. Seebeck coefficient of copper-constantan couple at different temperatures

Temperature (K)	Seebeck Coefficient (μV /K)
300	-40.83
350	-45.02
400	-48.65
450	-51.8
500	-54.62
550	-57.07
600	-59.21
650	-61.16
700	-61.77

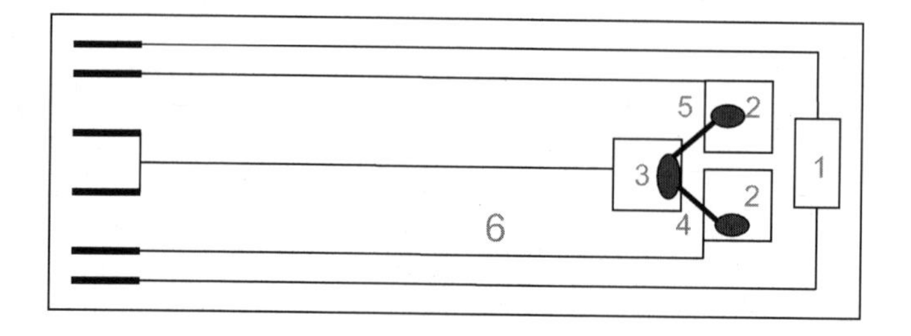

Fig. 1A Schematic of Seebeck Coefficient measurement system. 1 – heater, 2, heat source, 3 heat sink, 4 reference constantan wire, 5 sample

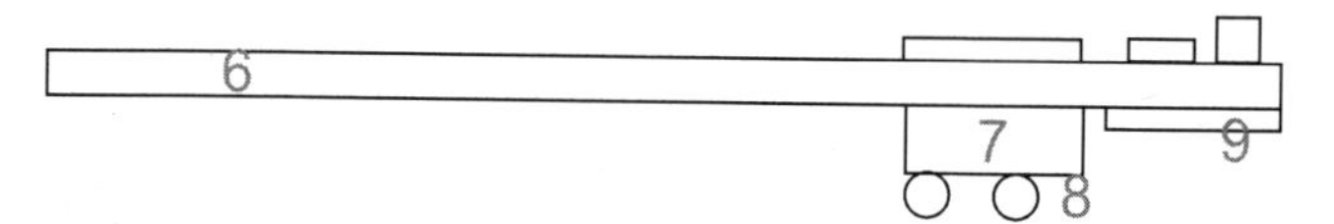

Fig. 1B 6 – Alumina substrate, 7 – copper block, 8 – hot stage heater, 9 – copper block

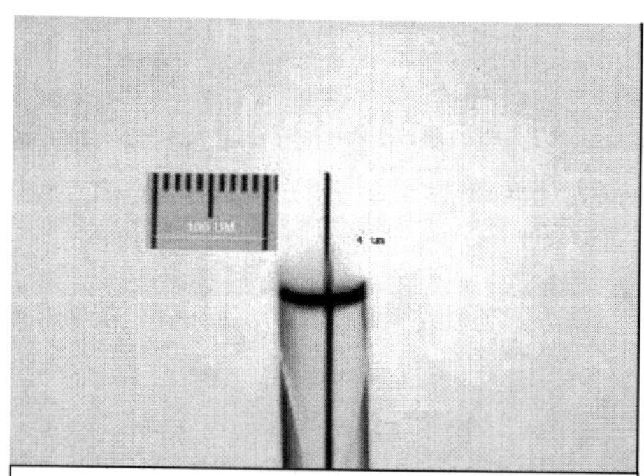

Fig. 2 Glass clad CdTe wire etched in HF for 5 minutes to expose the semiconductor core

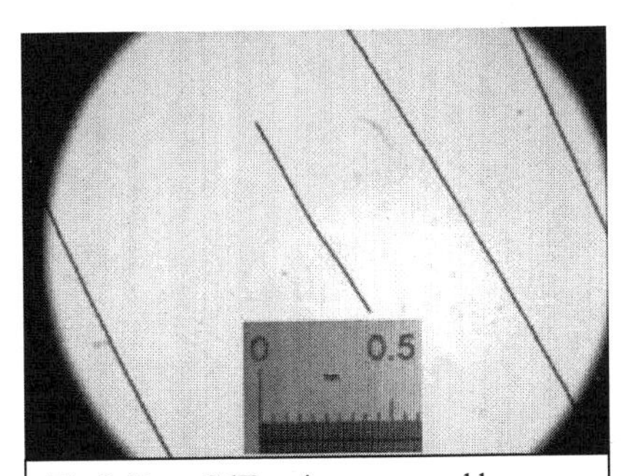

Fig 3. Bare CdTe wires prepared by etching the glass cladding completely.

Fig 4. Thermopower measurement setup for a fully bare 6 μm CdTe diameter sample.

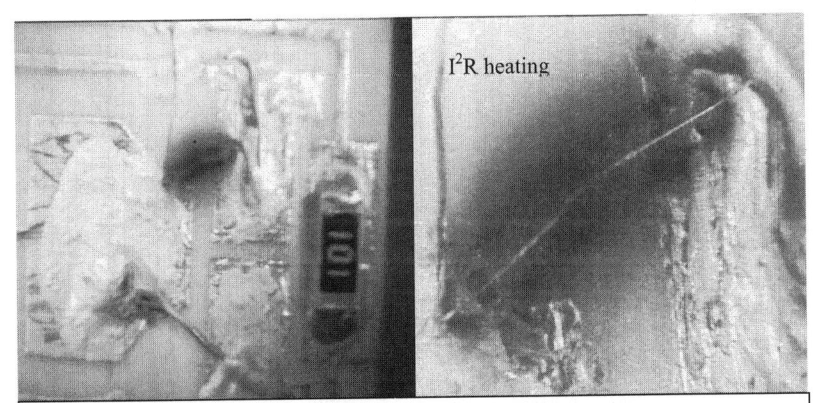

Fig 5 A – B. Pictures showing joule heating and the resultant damage to the CdTe wire upon passing 0.16mA current through the 10 μm sample.

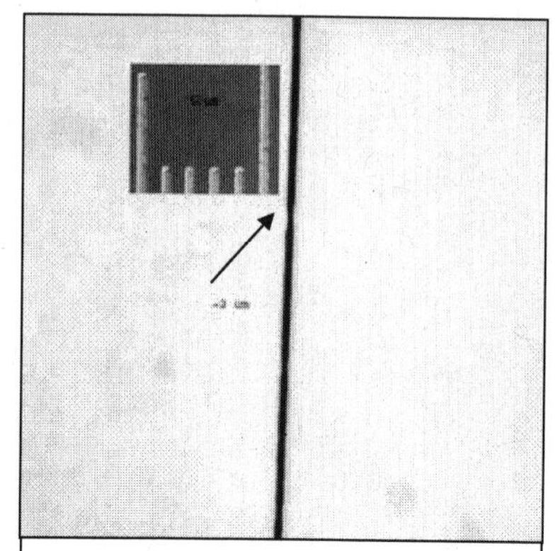

Fig 6. A 6 μm CdTe wire with a 2 μm μm constriction

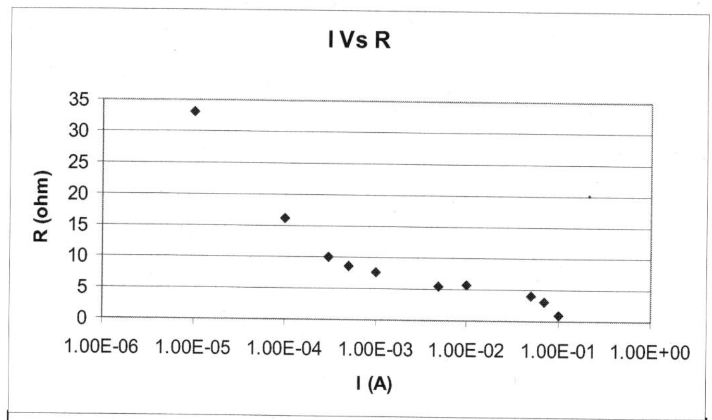

Fig.7 The variation of resistance as a function current in a typical semiconductor MNW (CdTe) showing both the ohmic and non-ohmic regions

STRUCTURAL CHANGES AND STABILITY OF PORE MORPHOLOGIES OF A POROUS GLASS AT ELEVATED TEMPERATURES

Brian Scott and Gary Pickrell
Materials Science and Engineering Department
Virginia Tech
Blacksburg, Virginia, USA

ABSTRACT

A spinodally phase separable porous glass was evaluated for pore morphology changes after undergoing heat treatments from 600°C to 900 °C. Glass samples with varying starting pore structures were treated at elevated temperatures for times of 25 minutes and 24 hours. The glass was characterized through the use of a nitrogen adsorption technique to determine the pore size distribution, surface area and pore volumes as a function of treatment conditions. Results show that some of the glasses have an average pore size that does not change with treatment conditions while other glasses show an evolution in the average pore size and pore morphology. The pore morphology of this can be controlled with heat treatments to the porous glass.

INTRODUCTION

Porous glasses derived through selective leaching of a spinodally phase separated glass have been produced for many years. The primary porous glass is produced from a sodium borosilicate glass and is sold under the Corning brand name of porous Vycor. Glass of this type has been used in many applications from separation membranes to the cladding in an optical fiber gas sensor[1, 2]. Typically the skeleton of the porous glass is a high silica composition with a high softening point. The continuous service temperature for the commercial version of the porous glass is 600°C, as given by Corning's product literature. It is of interest to understand how the pore structure evolves over time at temperatures above the given continuous service temperature. The evolution of the pore structure may affect the transport properties of material through the glass and the optical properties of the porous glass.

Originally the production of porous glass was an intermediate step in the production of a high silica glass. The solid glass article is heated to above 900 °C to fully consolidate the pore structure, thus producing a high silica glass [3-5]. At temperatures above this, the glass has a low enough viscosity for the pores to collapse where the main driving force is the reduction in the surface energy that is associated with the surface area of the pores [6]. A mathematical model has been developed that describes the consolidation of a porous glass where all of the pores in the glass have the same pore width. Rates of sintering proposed are a function of viscosity of the glass structure. The models cover sintering and consolidation of optical fiber preforms, sol-gel structures and porous glasses with the model being supported by some experimental evidence [7-9]. All of the thermal processing for the experimental verification was done at temperatures above 1000°C where the viscosity of the glass is sufficiently low to allow for viscous flow. The same author also developed a model to describe consolidation of porous glass that is in a constrained state[10]. The example given was of a silica sol-gel contained within a fiber mat, where the mat restricted the consolidation of the sol-gel. The modeling for a constrained environment describes a coarsening of the pores in which smaller pores become part of the larger pores in the material. The driving force is proposed to be the balancing of the reduction of surface energy with the strain energy in the material [10]. Several studies have been done with a thermal treatment of non alkaline treated porous glasses at temperatures less than 900°C. These studies indicated that coarsening occurs.[11-14] However, these studies do not track the evolution of the pore structure as a function of treatment time, and only report the change in the average pore size as a function of temperature for static times.

MATERIALS AND METHODS

Bulk glass samples were prepared by crushing a non phase separated Vycor glass tubing produced by Corning, Inc. The glass was crushed into pieces that were approximately 5mm square. These glass pieces were loaded into an alumina crucible and placed into a furnace that was already at operating temperature. Before insertion of the crucible, the furnace temperature at the location where the crucible would be placed was probed with a thermocouple to ensure the appropriate temperature had been reached at that location. After adjusting the furnace to the appropriate temperature, the crucible filled with the green glass was inserted and heat treated for 20 hours at 550 °C and then allowed to cool at the unpowered furnace cooling rate.
Once the crucible was cool, the glass was removed and leached. The leaching process started with the removal of the surface layer by immersion of the glass in a 5% ammonium biflouride solution for 10 minutes followed by immediate immersion into a beaker of 3N HNO_3. The nitric acid solution was heated to 90°C and left for a period of two days to ensure complete removal of the secondary phase. During the leaching process the top of the beaker was covered to reduce evaporation of the solution and any solution that did evaporate was replaced on a daily basis. After two days the glass was removed from the acid and rinsed in DI water for two more days and then air dried. The water was replaced daily during the rinsing process.

After the initial leaching process the glass pieces were treated with a 0.5 N NaOH solution and then post heat treated. Treatment of the glass with NaOH was done to remove any secondary silica in the pores. In removing the silica gel from the pore structure, any consolidation of the gel should not be present in the resultant data. The NaOH treatment of the glass pieces was done for a period of 24.5 hours. This was done to adjust for the greater diffusion distance that is present in the porous glass samples. The cladding in the previously mentioned optical fiber has a outside diameter of approximately 200 μm, so the diffusion distance is a factor of 7 greater in the porous glass. Treatment time for the NaOH immersion was determined by assuming that the diffusion condition of the NaOH was the same for both the fibers as for the bulk glass pieces. If the conditions are the same then the diffusion distance is the only adjustment that needs to be taken into account. Since the NaOH treatment involves the dissolution of the silica gel in the pore structure and its transport out of the glass, then the diffusion distance affects the time needed for the treatment. Diffusion of the dissolved silica gel out of the pore structure and diffusion of the NaOH solution into the pores will be affected by the distance for these species need to diffuse. A greater distance will require a longer time for complete dissolution and removal of the silica gel if all other diffusion related factors are the same. Time in solution was calculated as follows with D being the diffusion distance required, t is the time in solution and α, a term which includes all other factors effecting diffusion. Diffusion in the fibers is denoted by an f subscript and in the bulk by b. Glass thickness is 1.4 mm and average fiber diameter is 200 μm.

$$D_f = \alpha\sqrt{t_f} \qquad D_b = \alpha\sqrt{t_b}$$

$$D_b = 7\ D_f = 7\alpha\sqrt{t_f} = \alpha\sqrt{t_b}$$

$$t_b = 49\ t_f = 49*30\text{min}$$

$$t_b = 1470\text{min} = 24.5\ \text{hours}$$

Once the bulk glass was treated with NaOH it was removed from solution and immersed in a solution of 3 N HNO_3 for several days to ensure complete neutralization and removal of the NaOH in the pore structure. The acid solution was removed by rinsing in DI water for several more days and then air drying prior to post heat treatment. Glass samples were then given a post heat treatment for times of 25 minutes and 24 hours at 600-900°C in 100°C increments.

Characterization

Samples were characterized by a nitrogen adsorption method to determine the pore size distribution, pore volume and surface area. The glass samples were placed into a sample holder and outgassed at 300°C for a period of at least 24 hours in order to ensure that all water is removed from the glass sample before testing. Sample weights are taken before and after outgassing to determine the dry sample weight being tested.

During the nitrogen adsorption test, 10 adsorption and 10 desorption points were used to produce an experimental isotherm. The isotherm was then used to determine the pore structure by duplicating the isotherm with a Monte-Carlo simulation. This modeling approach to produce usable data is done by determining what the isotherm for a particular pore size and pore shape would be and then adding up the isotherms to reproduce the experimental isotherm. The range of pore size for this type of analysis is from 2.5 nm up to 77 nm with the full width of the distribution being 75.19 nm. The output from this is a set of data that includes the cumulative pore volume, the cumulative surface area, and the derivative of both the cumulative pore volume function and cumulative surface area function. The derivatives give the pore size distributions as a function of the volume or surface area of the pore size.

The current analysis was done using a Monte-Carlo model of the adsorption isotherm based on a cylindrical pore shape with nitrogen being adsorbed onto a silica surface. Data that was output was then processed to produce a pore volume and the percent of total pore volume for each pore size. This data conversion allowed for the determination of the average pore size as function of pore volume.

Samples were also characterized through scanning electron microscopy. Images of the samples were taken from fresh fracture surfaces. The micrographs are of fracture surfaces of the glass samples taken near the center portion of the glass piece to reduce any images due to near surface etching that might occur during the leaching process. The center portion is thought to most closely represent the pore structure that would be near the core of the porous clad fibers. Micrographs were taken with an acceleration voltage of 5Kv and a magnification of approximately 44Kx.

RESULTS AND DISCUSSION

The data that was accumulated during the analysis of these glass samples included the adsorption and desorption isotherms, the output from the Monte-Carlo analysis and the data that was derived from the Monte-Carlo outputs. The data presented is the average pore size, pore volume, and surface area.

Figure 1 and Figure 2 show the change in pore structure as a function of time for the porous glass samples that are heat treated at different temperatures. Figure 1 combines the average pore width and the pore volume of the glass sample. Figure 2 details the changes in the total surface of the porous glass samples. During the extended heat treatment none of the samples have a more than a 50% decrease in pore volume while the pore width and surface area have a wide range of values.

From the values in Figure 1 and Figure 2, several observations can be made. In the 600°C samples the pore volume decreases for both 25 minute and 24 hour samples, while the average pore width changes very little for the 25 minute sample and decreases by 11.5%. The 700°C sample shows little change in the average pore width, but the pore volume and surface area for the samples decreases at all heat treatment times. In the 800°C sample group the change in the average pore width is one of overall increase at treatment times of 25 minutes and 24 hours. The pore volume initially increases and then decreases at the longer heat treatment time, while the surface area decreases at all treatment times. At a treatment temperature of 900°C the sample has a slight reduction in the average pore width at 25 minutes and a large reduction in the sample treated for 24 hours. The surface area for the sample has an initial decline followed by a substantial increase in the 24 hour sample, while the pore volume decreases at all treatment times.

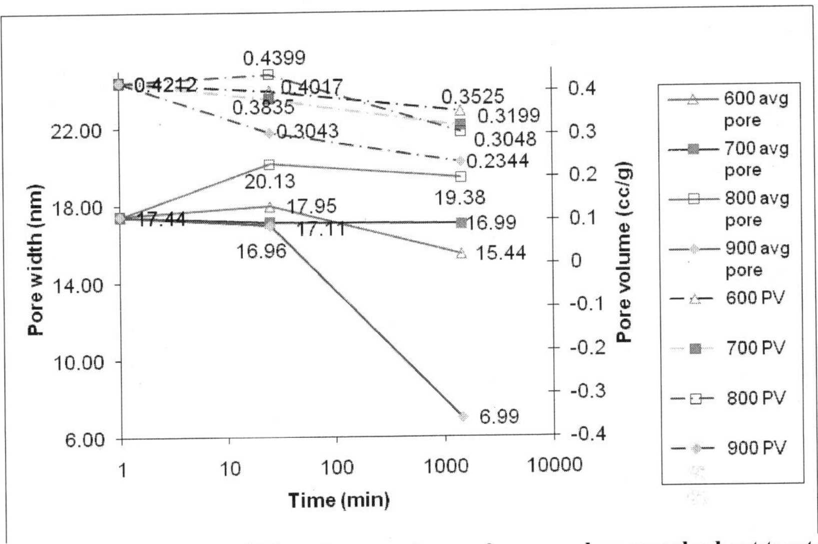

Figure 1. Average pore width and pore volume of porous glass samples heat treated above 600 °C

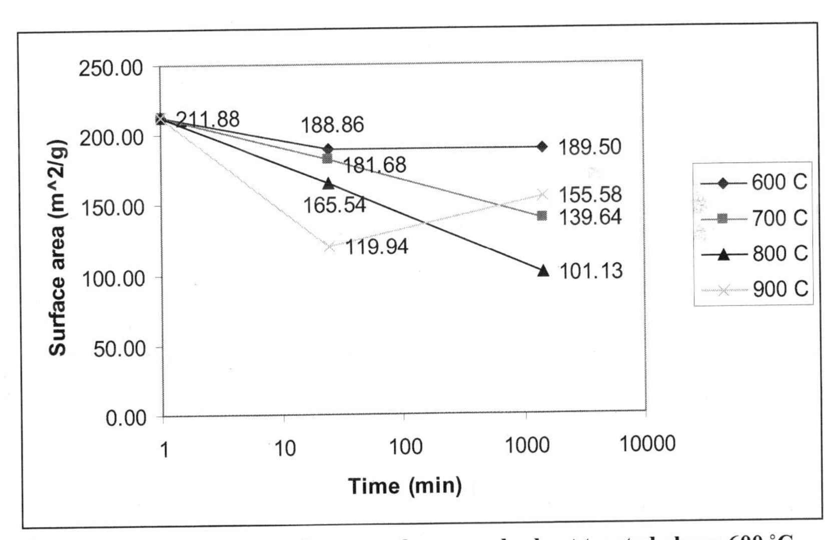

Figure 2. Surface area of porous glass samples heat treated above 600 °C

Table I details the changes in the pore characteristics based on the heat treatment conditions. Results are reported as percent change from the initial state and from any shorter heat treatment time. The table shows trends of change in the pore characteristics based on the thermal processing. Samples heat treated at 600°C have an initial increase in pore width with an accompanied decrease in pore volume and surface area. Changes in the group are small with the exception of the surface area which has a reduction of about 11%. With increase treatment time, the average pore width and pore volume have a significant decrease while the surface area remains relatively unchanged. In the 700°C sample group, the average pore width does not change much with any treatment time while the pore volume and surface area show a significant continual decrease. Pore width and pore volume increase in the 800°C samples treated for 25 minutes with an accompanied surface area decrease. When treated for longer times the 800°C samples have a slight decrease in pore width with a relatively large decrease in surface area and pore volume. Samples in the 900°C treatment group have an overall large reduction in pore width, pore volume and surface area as the treatment time increases.

Table I. Percent change in average pore width, pore volume and surface area for heat treated porous glass samples

		Percent change from previous treatment time			Percent change from initial condition		
Sample Heat treatment temperature °C	Heat treatment time (minutes)	Average pore width (nm)	Pore volume (cc/g)	Surface area (m^2/g)	Average pore width (nm)	Pore volume (cc/g)	Surface area (m^2/g)
600	25	+2.9	-4.6	-10.9	+2.9	-4.6	-10.9
600	1440	-14	-12.2	+0.33	-11.5	-16.3	-10.6
700	25	-1.9	-8.9	-14.2	-1.9	-8.9	-14.2
700	1440	-0.7	-16.6	-23.1	-2.6	-24.0	-34.1
800	25	+15.4	+4.4	-21.9	+15.4	+4.4	-21.9
800	1440	-3.7	-30.7	-38.9	+11.1	-27.6	-52.3
900	25	-2.75	-27.8	-43.4	-2.75	-27.8	-43.4
900	1440	-58.7	-23.0	+29.7	59.9	-44.3	-26.6

Scanning electron micrographs of fracture surfaces of the samples are shown in Figure 3 through Figure 11. Inspection of the micrographs shows the type of pore shape and overall appearance of porous nature of the samples. The pore shape can be described as cylindrical pores or a series of overlapping cylindrical pores that produce a more slit like pore shape. Figure 3 is an SEM micrograph of a piece of porous glass from the base group that was produced prior to any further thermal processing. The micrographs in Figure 4 through Figure 11 are micrographs of all of the thermally treated samples and can be compared to Figure 3 to visually determine if any changes take place during the thermal processing. Most of the images taken of the fracture surfaces do not visually reflect the changes shown in the other data. This is due to the small area imaged, the surface roughness of the sample, and the relatively small changes in overall pore structure. While these micrographs might not reflect the changes that take place as described by the nitrogen adsorption data, they do show that the pore shape remains unchanged during the thermal processing.

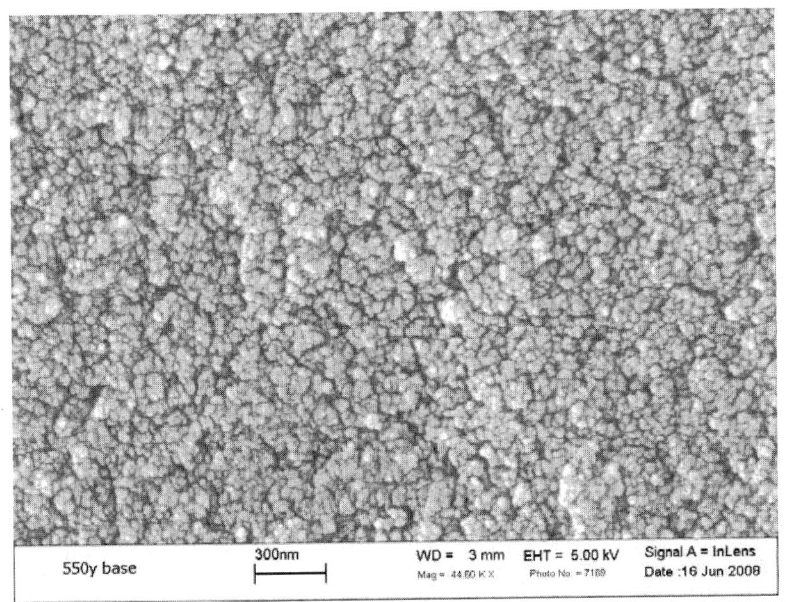

Figure 3. SEM micrograph of fracture surface for porous glass sample with no heat treatment

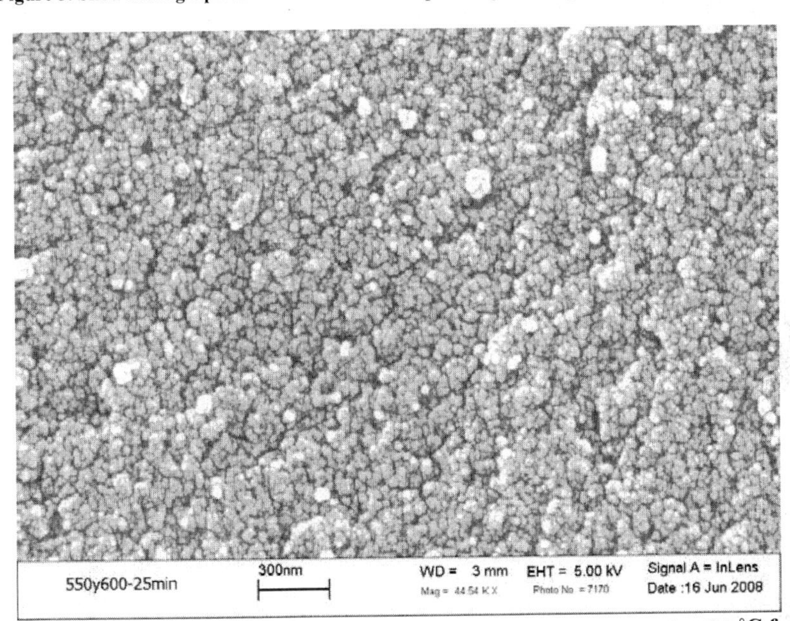

Figure 4. SEM micrograph of fracture surface for porous glass sample heat treated at 600 °C for 25 minutes

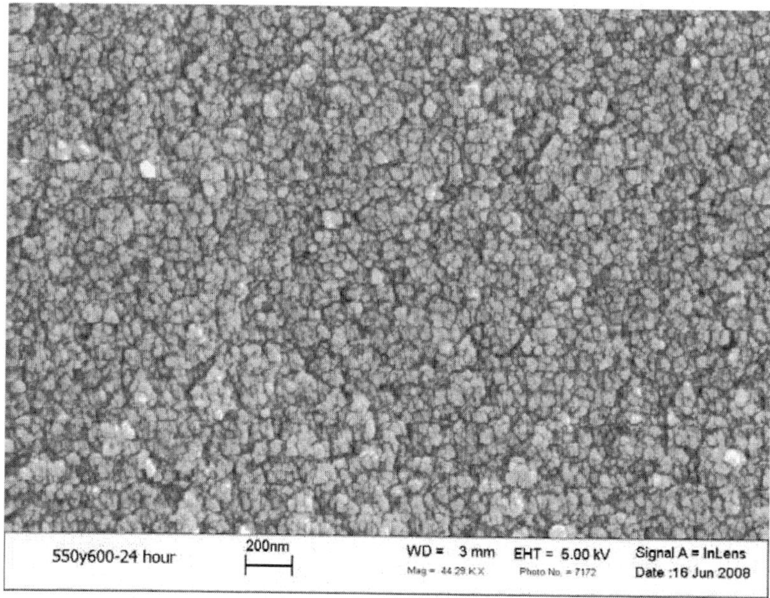

Figure 5. SEM micrograph of fracture surface for porous glass sample heat treated at 600 °C for 24 hours

Figure 6. SEM micrograph of fracture surface for porous glass sample heat treated at 700 °C for 25 minutes

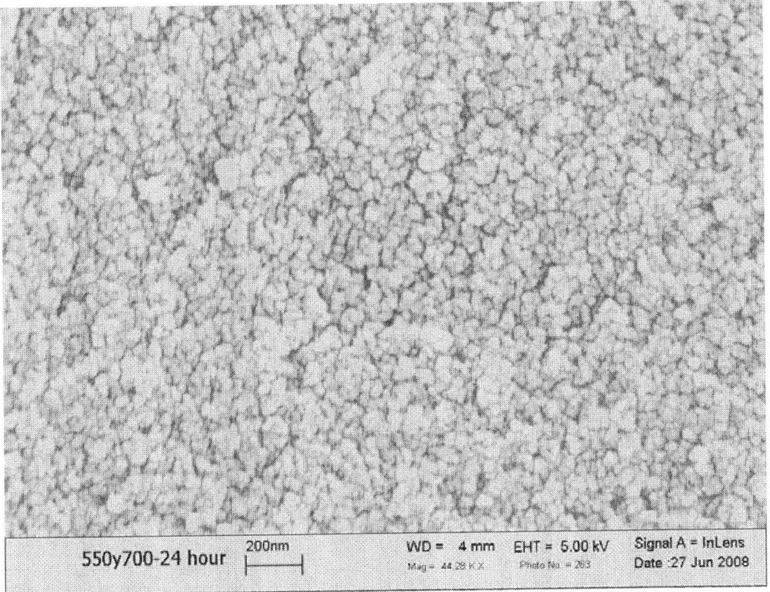

Figure 7. SEM micrograph of fracture surface for porous glass sample heat treated at 700 °C for 24 hours

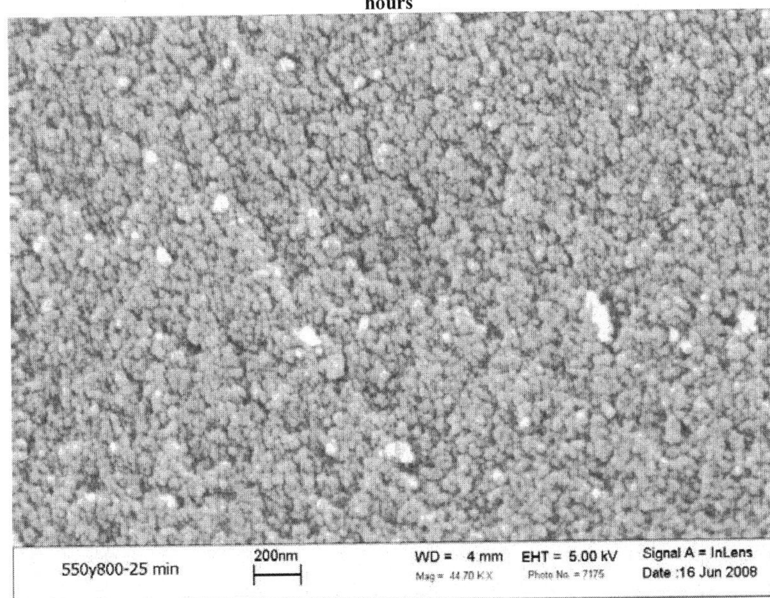

Figure 8. SEM micrograph of fracture surface for porous glass sample heat treated at 800 °C for 25 minutes

Figure 9. SEM micrograph of fracture surface for porous glass sample heat treated at 800 °C for 24 hours

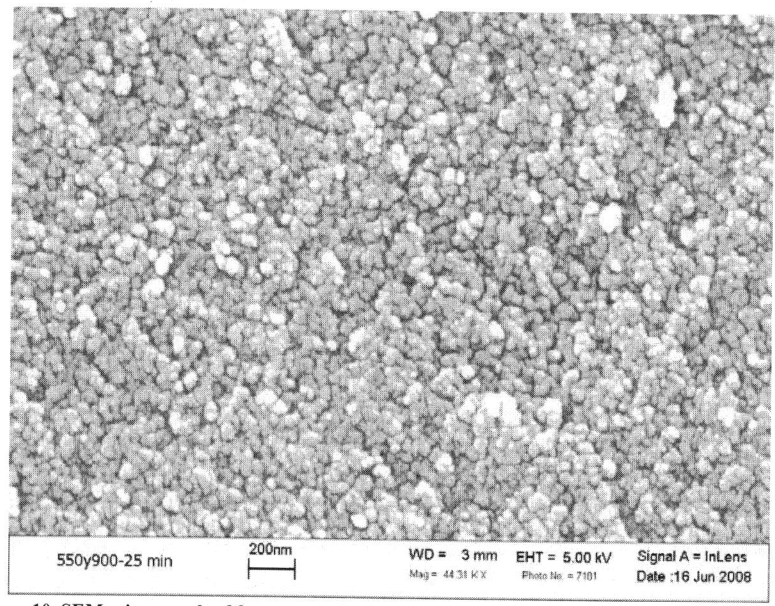

Figure 10. SEM micrograph of fracture surface for porous glass sample heat treated at 900 °C for 25 minutes

Figure 11. SEM micrograph of fracture surface for porous glass sample heat treated at 900 °C for 24 hours

The objective of the experiment was to see how the pore characteristics of a porous glass used as a cladding material in an optical fiber would change when exposed to temperatures above 600°C. Samples were characterized using a nitrogen adsorption technique and pore shape was verified through SEM images of the samples. Analysis of the nitrogen adsorption data assumes a cylindrical pore shape and the SEM images are in agreement with this assumption. The nitrogen adsorption data was reduced to show the average pore width, pore volume and surface area for the samples. Pore characteristic data shows trends that indicate the type of evolution in the morphology of the glass.

Assuming that the basic pore shape of the glass does not change, the way that the average pore width, pore volume and surface area changes narrows down the types of changes in pore structure that are possible. When the average pore width increases and both the pore volume and surface area decrease, then the change is likely due to the consolidation of smaller pores. Through consolidation of the small pores in the glass, the average pore width will be increased as there will be fewer small pores and the overall pore volume and surface area will decrease as the small pores are eliminated. When the average pore width does not change, but the pore volume and surface area both decrease then both large and small pores are consolidating. However, if the average pore width increases, the pore volume stays the same or decreases and the surface area decreases, then this is mostly likely due the coarsening of small pores into larger pores. This trend will come about as the overall reduction in the amount of small pores with the growth of the larger pores will shift the average pore width up. In the combining of smaller pores to form larger pores of the same approximate volume, the surface area will decrease as larger pores of the same volume will have less total surface area than the equivalent volume of the smaller pores.

Table I details the changes in the pore characteristics of the samples treated for various times and temperatures. In the table, the trends of the pore changes are apparent and from these trends some conclusions can be made. Conclusions are based on the assumption that the basic pore shape does not change during the heat treatment process. This assumption is backed up by the SEM images is Figure 3 through Figure 11. The SEM images show cylindrical pores that overlap to form a longer slit like pore. The pore shape does not change in the any of the SEM images as the temperature or time of the treatment is varied.

In looking at the trends from the graphs and Table I, the types of changes in the pore structure can be deduced. The 600°C sample first has a slight pore width increase and a decrease in both the pore volume and surface area. This would indicate that the smaller pores in the sample are consolidating. With further thermal processing, the 600°C sample has a decrease in the average pore width and pore volume with very little change in the surface area. Morphological changes of this type would occur if the surface area is being created during the consolidation of pores within the material to compensate for the reduction during the consolidation process. A mechanism for this type of behavior is not known and the stable surface area during the 24 hour treatment at 600°C is considered a possible point of experimental error. Trends in the 700°C treated group initially show consolidation of the large and small pores by a stable pore width and a decrease in pore volume and surface area. Further heating continues the consolidation of the large and small pores. The 800°C sample data indicates that during the beginning heat treatment, the small pores coarsen to form larger pores. This is evidenced by the increased average pore width accompanied by a decrease in pore volume and surface area. Samples heat treated for 24 hours have a small reduction in the pore width and a decrease in pore volume and surface area, which signifies that a greater percentage of large pores are consolidating in the glass. In the 900°C samples for both times, consolidation of all of the pore sizes is apparent by the decrease in pore width and pore volume. However in the 24 hour treated group there is an increase in surface area. This could be experimental error or caused by the generation of more surface area in the manner in which the large pores consolidate. If during the consolidation, walls are formed as part of the pore closure process in the existing pores this would generated added surface area while reducing the average pore width and pore volume in the sample.

SUMMARY

A spinodally phase separated glass was leached, treated with sodium hydroxide to remove any secondary silica gel, and further thermally processed to determine the structural changes to the glass. All samples experienced morphological change during heat treatment of 25 minutes and 24 hours. None of the samples exhibited pore structural stability within the temperature range of 600°C to 900°C. The highest rate of change for all samples was within 25 minutes and is significantly less during the remaining part of the 24 hour heat treatment. Pore consolidation was evident in all of the samples with the exception of the 800°C sample treated for 25 minutes, in which pore coarsening is present.

Acknowledgement: The financial support of the US Department of Energy's National Energy Technology Laboratory (NETL) under award number DEFC2605NT42441 is gratefully acknowledged.

Reference:

[1]T.H. Elmer, "Porous and reconstructed glass," pp. 427-432 in Engineered Materials Handbook Vol. 4: Ceramic and Glasses. Edited. ASM International, 1992.
[2]B. Scott, et al., "Novel Chemical and Biological Fiber Optic Sensor," p. 66560F in Micro(MEMS) and Nanotechnologies for Defense and Security, Vol. 6556. Edited by T. George and Z. Cheng. Proc. SPIE 6556, Orlando, Florida, 2007.
[3]H.P. Hood and M.E. Nordberg, "Borosilicate Glass ", Vol. 2,221,709, United States 1940.
[4]H.P. Hood and M.E. Nordberg, "Method of Treating Borosilicate Glasses," Vol. 2,286,275, United States, 1942.
[5]H.P. Hood and M.E. Nordberg, "Treated Borosilicate Glass," Vol. 2,106,744, United States, 1938.
[6]T.H. Elmer, "Sintering of porous glass," *American Ceramic Society Bulletin*, 62[4] 513 516.
[7]G.W. Scherer, "Sintering of Low-Density Glasses: I, theory," *Journal of the American Ceramic Society*, 60[5] 236-239 (1977).
[8]G.W. Scherer, "Sintering of Low-Density Glasses: III, Effect of a Distribution of Pore Sizes," *Journal of the American Ceramic Society*, 60[5] 243-246 (1977).
[9]G.W. Scherer and D.L. Bachman, "Sintering of Low-Density Glasses: II, Experimental Study," *Journal of the American Ceramic Society*, 60[5] 239-243 (1977).
[10]G.W. Scherer, "Coarsening in a Viscous Matrix," *Journal of the American Ceramic Society*, 81[1] 49-54 (1998).
[11]T.V. antropova, et al., "Structural Transformations in Thermally Modified Porous Glasses," *Glass Physics and Chemistry*, 33[2] 109-121 (2007).
[12]I. Drozdova, T. Vasilevskaya, and T. Antropova, "Structural Transformation of secondary silica inside the porous glasses according to electron microscopy and small angle x-ray scattering " *Physics and Chemistry of Glasses: European Journal of Glass Science and Technology Part B*, 48[3] 142-146 (2007).
[13]T. Hirata, s. Sato, and K. Nagai, "Effects of Thermal Treatment on Gas Transport Through Porous Silica Membranes," *Separation Science and technology*, 40 2819-2839 (2005).
[14]A.V. Volkova, et al., "The Effect of Thermal Treatment on the Structural and Electrokinetic Properties of Porous Glass Membranes," *colloid Journal*, 67[3] 263-270 (2005).

Author Index

Annamalai, S., 135
Arai, K., 109

Battogtokh, J., 135
Bhatta, R., 135
Biering, I., 71
Bunker, R. S., 3

Chao, S., 23

Dogan, F., 23
Dutta, B., 135

Endo, Y., 41

Fukui, S., 31
Fukui, T., 31

Honjo, T., 99

Ito, M., 41, 51, 59

Kahvecioglu, O., 77
Kodera, Y., 31
Kondo, W., 109
Kumagai, T., 109
Kurosaki, K., 41, 59

Manabe, T., 109
Menon, M., 71
Miki, J., 31
Muta, H., 41, 51, 59

Nishioka, S., 51

Ohyanagi, M., 31

Pegg, 135
Pfeiffenberger, N. T., 117
Pickrell, G. R., 117, 123, 145
Pryds, N., 71

Rao, M. L., 87
Roy, R., 87

Scott, B., 123, 145
Shiozaki, S., 31
Sohma, M., 109

Timur, S., 77
Tsukada, K., 109

Uno, M., 41, 51, 59, 99

Wooddell, M. G., 123

Yamaguchi, I., 109
Yamanaka, S., 41, 51, 59, 99
Yamasaki, H., 109
Yamasaki, N., 31
Yin, J., 31